Illuminating Instruments

Illuminating Instruments

artefacts

STUDIES IN THE HISTORY OF
SCIENCE AND TECHNOLOGY,
VOLUME 7

EDITED BY

Peter J. T. Morris and Klaus Staubermann

MANAGING EDITOR

Martin Collins, Smithsonian Institution

SERIES EDITORS

Robert Bud, Science Museum, London

Bernard Finn, Smithsonian Institution

Helmuth Trischler, Deutsches Museum, Munich

Published in cooperation with ROWMAN & LITTLEFIELD PUBLISHERS, INC.

Smithsonian Institution
Scholarly Press

WASHINGTON, D.C.

2010

The series "Artefacts: Studies in the History of Science and Technology" was established in 1996 under joint sponsorship by the Deutsches Museum (Munich), the Science Museum (London), and the Smithsonian Institution (Washington, DC). Subsequent sponsoring museums include: Canada Science and Technology Museum, Istituto e Museo Nazionale di Storia della Scienza, Medicinsk Museion Kobenhavns Universitet, MIT Museum, Musée des Arts et Métiers, Museum Boerhaave, Národní Technické Museum, Prague, National Museums of Scotland, Norsk Teknisk Museum, Országos Mıszaki Múzeum Tanulmánytára (Hungarian Museum for S&T), Technisches Museum Wien, Tekniska Museet–Stockholm, The Bakken, Whipple Museum of the History of Science.

Published by SMITHSONIAN INSTITUTION SCHOLARLY PRESS
P.O. Box 37012, MRC 957
Washington, D.C. 20013-7012
www.scholarlypress.si.edu

In cooperation with
ROWMAN & LITTLEFIELD PUBLISHERS, INC.
A wholly owned subsidiary of The Rowman & Littlefield Publishing Group, Inc.
4501 Forbes Boulevard, Suite 200, Lanham, Maryland 20706
www.rowmanlittlefield.com
Estover Road
Plymouth PL6 7PY
United Kingdom

Front cover image: Dollond's table air pump, ca. 1780–1820. Courtesy Science Museum, Science & Society Picture Library.

Back cover images: (left) Kaufmann apparatus for measuring electron mass, 1906. Courtesy Deutsches Museum, Munich. (right) Ladd's electric egg, ca. 1850. Courtesy Royal Institution, Science & Society Picture Library.

British Library Cataloguing in Publication Information Available

Library of Congress Cataloging-in-Publication Data:
Illuminating instruments / edited by Peter Morris and Klaus Staubermann.
    p. cm. — (Artefacts : studies in the history of science and technology ; v. 7)
    Includes bibliographical references and index.
    ISBN 978-0-9788460-3-9 (cloth : alk. paper)
    1. Scientific apparatus and instruments—Congresses. 2. Science and civilization—Congresses. 3. Science—History—Congresses. 4. Technology—History—Congresses. I. Morris, Peter John Turnbull. II. Staubermann, K. B. (Klaus B.)
    Q184.I45 2010
    681'.75—dc22   2009027953

ISBN-13: 978-0-9788460-3-9
ISBN-10: 0-9788460-3-6

Printed in the United States of America

♾ ™ The paper used in this publication meets the minimum requirements of the American National Standard for Permanence of Paper for Printed Library Materials Z39.48–1992.

Contents

■ V

Series Preface

IN RECENT YEARS, as more and more historians of science and technology have found employment in museums, there has been a growing concern that objects have not received proper attention in serious historical studies, and even in exhibits they more often serve as icons than as historical evidence. Yet there has been no concerted effort to resolve this dilemma. There is little relevant training in academic programs, and museum scholars have failed to develop a compelling body of literature that might serve as a model for themselves and their academic colleagues.

Beginning in the fall of 1993, the three of us, with encouragement from colleagues in both academic and museum worlds, considered what we might do to address this problem. With help from our respective institutions, we established "Artefacts," a name associated on the one hand with annual meetings that focused on different topics where these issues might be discussed, and on the other hand with a series of books with articles based on these same topics.

With this volume, the seventh, we feel that we have reached a milestone. The Smithsonian Institution Scholarly Press has taken over responsibility for publishing from the Science Museum, with whom we have had a very happy relationship. We have taken advantage of the time it has taken to make the change to give ourselves a more formal structure, notably to establish an Advisory Editorial Board, listed on the copyright page. This should strengthen our intellectual base and—most important—provide volume editors with welcome support.

At the same time we are making formal recognition of what we hoped would be true from the start: that this would be a collaborative effort, involving as many museums dealing with the history of science and technology as possible. Our only criteria for inclusion being an expression of support for our goals and the commitment of a very modest amount of staff time, as appropriate, in their pursuit.

The expansion of our list of sponsoring museums, to date, is noted on the copyright page.

Thanks to all of those institutions and people who have leant their support; we look forward to collaboration with many more.

Robert Bud, Science Museum, London
Bernard Finn, Smithsonian Institution
Helmuth Trischler, Deutsches Museum, Munich

Introduction

THIS SEVENTH VOLUME in the Artefacts series, which began in 1999, deals with the artefacts most closely identified with science museums and, indeed, science, namely scientific instruments. It has its origins in an Artefacts meeting held in October 2004 at the University Museum in Utrecht titled "Scientific Instruments as Artefacts: Shiny Objects and Black Boxes," which was cosponsored, like all previous Artefacts conferences, by the Deutsches Museum, Science Museum, and Smithsonian Institution and organized by one of the editors of this volume (Klaus Staubermann) who was then a curator at the University Museum. Most of the papers were originally presented at this meeting, one (by Peter Morris) was originally presented at an earlier Artefacts meeting about the environment held in Munich in 2000, and two others (by Deborah Jean Warner and Sean Johnston) were specially commissioned for this volume.

The definition of a scientific instrument has been a matter of some debate in recent years. Deborah Warner has shown how in the nineteenth century instruments became "scientific." Before that, they would have been either "mathematical" or "philosophical" instruments, depending on the ways they were employed. In the early twenty-first century scientific instruments are considered as "research technology." Regardless of how one defines a scientific instrument, and we have taken a pragmatic view in this volume, they illuminate many things. Of course some scientific instruments, including two in this volume—the solar microscope and the magic lantern—provide illumination in a direct sense but all scientific instruments shed light on scientific phenomena. Furthermore, they are the product of local skills and practices, upon which this volume also sheds light. But the illuminating role of scientific instruments goes much further than that. We would argue that they can be sensitive probes for revealing the scientific, social, and cultural environment of the instrument even after it has ceased to be used actively as an instrument, which the chapters in this volume demonstrate. The volume shows how scientific instruments are an excellent probe for local culture and cultural contexts. Klaus Staubermann, in his reworking of demonstration practice, has come to a similar conclusion: instruments are an expression of their local culture as well as rendering this local culture by their very use. Peter Heering's chapter shows how the reconstruction of solar microscopes sheds light on the various uses of these instruments and the nature of scientific dilettantism in the eighteenth century. By contrast Deborah Warner's chapter illuminates the use of instruments for quality control in industry and the search for new ways of testing samples, thus revealing the nature of nineteenth-century industrial analysis. In a later period, Peter Morris's chapter on the analysis of DDT demonstrates how concern about pesticides residues increased in the 1950s well ahead of the publication of *Silent Spring* in 1962. Fittingly for a series concerned with the study of artifacts in museums, Christian Sichau's chapter explains how an instrument's environment within a museum changed over time. This reflects the arguments made by historians of science—for instance by Christoph Lüthy—that the perception of an instrument is highly dependant on the instruments' ability to perform its original functions within the museum.

Crucially, however, scientific instruments are not only directly useful to scientists and a cultural probe for historians of science, but also illuminate the evolution and practice of science to a general public. They reveal the cultural context of science and act as symbols for broader historical developments within the museum. The refractometer can be seen as an example for the development of quality control here, the electron capture detector for the growth of environmental concern, CFCs as well as DDT, and the magic lantern for the changing nature of popular science. It is the task of historians to extract the different meanings of scientific instruments and place them in their broader context, and it is the duty of museum curators to weave these different layers of meaning into compelling stories that will both captivate and inform museum visitors. Sean Johnston's chapter reveals the problems of trying to display a subject, holography, whose historical meaning and popular appeal changes over time, with implications for our understanding of holography's development and the problems of collecting material relating to this protean subject. Jane Wess's chapter on a Science Museum exhibition, "Inside the Atom," illustrates the challenges of presenting the latest historiography of science and scientific instruments to the general public, showing how an expectation of scientific heroes and iconic objects can be carefully diverted toward a broader understanding of scientific developments. Finally, Arne Schirrmacher warns us that any attempt to present scientific instruments to a broader public should be governed by trying to make the artefact itself "visible" rather than content us with simply presenting attractive visualizations. As historians, we hope this volume will encourage more studies of scientific instruments as historical and cultural probes that illuminate both scientific practice and social change. As curators, we seek to promote the analytical study of scientific instruments in museum collections and the creation of new stories, allowing our visitors to draw their own meanings from these objects, which are central to the development of modern science.

Instruments
in Practice

CHAPTER 1

Instrument and Image

INSIDE THE NINETEENTH-CENTURY SCIENTIFIC LANTERN SLIDE

Klaus Staubermann

Introduction

Historians of science have frequently argued that the nineteenth century marked a shift from demonstration instruments toward pictorial representations in astronomy. Although demonstration instruments never disappeared from science teaching, pictorial representations, indeed, played a strong role from the late nineteenth century onward. In this chapter I argue that already in the beginning of the nineteenth century new astronomical discoveries led to an increasing demand of images of newly discovered objects. This chapter examines how demonstrators and projectionists met this demand by what was already an established medium at its time: magic lantern slides. Based on my experience with original historic slides I will show how these devices came into being and were used. Demonstration instruments, instead of being replaced by images, on the contrary, were at the very center of producing those images.

Images and Instruments

In 1879 Camille Flammarion, a trained astronomer and engraver, published one of the most extensively illustrated popular science books of the nineteenth and early twentieth centuries, *Astronomie Populaire* (see figure 1.1). His book was not only richly illustrated, but it also was a good value and within a short time became a bestseller in France and elsewhere. His numerous illustrations usually showed celestial objects, instruments, and astronomical diagrams.

The nineteenth century has often been portrayed as the age of visual representation. Although representations of the heavens are as old as the science of astronomy itself, it was in the nineteenth century that pictorial representations became prominent. The number of illustrations, and later photographs, increased significantly from the 1830s onward. A popular

FIGURE 1.1
Cover of Camille Flammarion's *Astronomie Populaire*, 1880. Courtesy Klaus Staubermann.

astronomy book equally richly illustrated and successful as Flammarion's book was Simon Newcomb's *Popular Astronomy*, which was translated in several languages and became a great success in Europe and elsewhere. In 1922 the seventh edition of *Populaere Astronomie* by Newcomb-Engelmann contained more than 240 images, many of them photographs.[1] Several other books could be mentioned here, including *Der Wunderbau des Weltalls* by Johann Heinrich Mädler or *Die Wunder des Himmels* by Johann Joseph von Littrow, published in nine editions between 1834 and 1910. Illustrations usually showed celestial objects, instruments, and astronomical diagrams.

The fashion of imaging in the nineteenth century was limited not only to books. Scientific journals would be as extensively illustrated as exhibition guides to museums or scientific institutions. For example, almost all double pages of the almost 100-page astronomy section of the guide to the newly founded Urania in Berlin were illustrated with either photographs or drawings of celestial objects, instruments, and so forth.[2] To meet this increased public demand publishers often would supply separate sets of astronomical images.[3] One can conclude that indeed in the second half of the nineteenth century and in the beginning of the twentieth century a large number of illustrations could be found in popular and teaching books of astronomy.

However, a closer look reveals that the story of illustrations is more complex than often argued. First of all, astronomy teaching books before the late or even mid-nineteenth century did contain illustrations already and often quite a few. During the eighteenth century, diagrams and engravings could be found on separate tables at the end of the book. For example, Johann Jacob Scheuchzer's introduction to astronomy, *Physica oder Naturwissenschaft*, published in two volumes in 1743 contained twenty-three illustrated tables. Johann Heinrich Voigt's *Lehrbuch der populaeren Sternkunde* of the same period contained two richly illustrated tables concerning spherical astronomy, and August Heinrich Christian Gelpke's *Allgemeinfaßliche Betrachtungen über das Weltgebäude* of 1801 contained diagrams and illustrations of lenses, telescopes, geometry, movement of Earth, solar eclipses, surface of the moon, the Milky Way, and so on. The published astronomical image has formed an integral part of astronomy books of the past three centuries and printed representations have drawn the attention of audiences since the invention of book printing.[4] On the other hand, not all popular astronomy books of the late nineteenth century were necessarily superbly illustrated: when in 1881 the self-nominated "translator of astronomical research to lay audiences" Adolph Drechsler published his *Illustriertes Lexikon der Astronomie*, the book hardly contained more illustrations than popular astronomy books published several decades before, despite its suggestive title.

So, what exactly happened in the nineteenth century and for what reasons? Indeed, there had been an increase of illustrations in astronomy books during the nineteenth century. Although this varied from book to book, the change from 1800 to 1900 was clearly visible. Images did become more prominent during the course of the nineteenth century. One of the main reasons for this change was technological progress. The historian of astronomy Klaus Hentschel mentions as possible factors contributing to this development the refinement of lithographic, engraving, and photographic practices in the nineteenth century.[5] Especially, the invention of photography played an important role for the establishment of high-quality images in the late nineteenth and early twentieth century. Furthermore, the industrialisation of printing processes led to a wider distribution of high-quality images throughout the nineteenth century.[6] Also, although printing techniques in fine arts and sciences could often be quite different, the new standards of the former affected the latter.[7]

The increase of illustrations in popular astronomy books was due not only to new techniques and technologies. The historian of science Susanne Utzt argues that the increase was also due to various new astronomical discoveries made during the nineteenth century.[8] Newly discovered celestial objects such as planetoids (or asteroids) as well as comets and finally, in September 1846, the discovery of the planet Neptune stimulated the imagination of the public. This in return led to an increasing demand for images of celestial objects as well as the instruments and astronomers connected to them.[9] Historian of science Klaus John argues that because of the dominance of images in the middle of the nineteenth century, demonstration instruments had begun to become obsolete.[10] In order to support his claim John quotes from *Meyers Grosses Conversations-Lexicon* of 1850:

> In general, the fondness of such machines has been lost very much; their expediency is indeed limited to first elementary education; with a fairly quick mathematical conception one comes to a much clearer conception of the celestial movements with illustrations and corresponding explanations than with all planetariums.[11]

The disappearance of demonstration instruments during the nineteenth century has been observed by many historians of science. However, explanations for their disappearance differ significantly. Historian Elly Dekker, for example, has argued that in the case of the armillary sphere their alleged disappearance was partly due to lighting technology:

> With the introduction of street lighting the boundary between day and night faded away. These technological developments of the nineteenth century did more for the downfall of the concept of the common Ptolemaic globe than did any revolution in science.[12]

But did astronomical demonstration devices disappear during the nineteenth century? And was their fate sealed by illustrations, indeed? I will show how in the nineteenth century popularizers promoted new discoveries in astronomy both by images and instruments. Before doing so I will briefly give an overview of astronomical demonstration devices until the beginning of this period.

Instruments and Their Alleged Disappearance

Historian of astronomy Ernst Zinner, in his opus magna on German and Dutch astronomical instruments from the eleventh to eighteenth centuries, dedicates an extensive chapter to teaching instruments. Zinner describes in detail the various types and kinds of instruments used for demonstrational purpose. He shows that from medieval times until the eighteenth century five main categories of instruments served as common demonstration devices: pointers, celestial globes, spheres, astrolabes, and clockworks.[13] Zinner also shows that often these devices could not be separated: for example, in 1563 the duke of Hesse, Wilhelm IV, obtained a device that was both celestial globe and clockwork. Also, combinations of celestial globes, spheres, and pointers were not rare.[14] It should be mentioned here that these were not the only astronomical demonstration devices at that time, only that these were the ones that usually could be found at courts and colleges.

The development of one kind of demonstration instrument Zinner describes in considerable detail is that of the armillary sphere. The armillary sphere consists of several rings, often from metal, that represent the most important celestial spheres: equator, zodiac, ecliptic, tropics, and so on. The armillary sphere enables the demonstrator to show, for example, how the seasons arise, why the length of the day depends on the time of year, and how the visibility of certain stars depends on their latitude.[15] Early armillary spheres would sometimes include markers or representations of the moon and sun, and some would even show the planetary spheres by means of rings.[16] Planetary instruments in the shape of movable disks had existed earlier, but when instrument makers equipped spheres and disks with clockworks the planetary clock was born. Many of these demonstration devices, dating from around the fourteenth century to the present day, can still be found. They prove that spectators were not only interested in grasping concepts of celestial motion, but also to see the pictorial representations of celestial bodies and their actual movements.

New and different attitudes toward representations of the sky were not limited to the nineteenth century.[17] Historian Yorck Alexander Haase, for example, argues that in the late eighteenth century, due to the increasing amount of detail, celestial globes began to be replaced by star charts.[18] Further, representations of stellar constellations changed frequently as well. Deborah Warner argues that changes of representations of scientific instruments on eighteenth century star charts represent actual changes of the instrumental design at that time.[19] Also, the uses of instruments changed with time. Whereas the armillary sphere served as a demonstration and observation device first, by the eighteenth century it had become a sole demonstration device—and slowly began to disappear from the instrument market.[20]

While demand for illustrations grew, demand for demonstration instruments and models grew too. When the German teacher C. W. E. Putsche published his work *Planetarium* in 1805 he made sure that it did not only contain text and images, but also a working model. Putsche argued that illustrations were not sufficient for teaching astronomy to the youth—models were more useful.[21] Nineteenth-century models and demonstration instruments came in various types, materials, and forms and would carry names such as lunarium, tellurium, and uranorama.[22] One such example is the cometarium by John Taylor. Taylor, in 1828, designed a device for showing the approximate positions of a comet. Although such mechanical cometariums were of rather accurate design, they were produced as orbital demonstration devices and were not intended to be predictive tools.[23]

Henry C. King, in his decisive study on the history of the planetarium, *Geared to the Stars*, shows how during the eighteenth and nineteenth centuries planetariums slowly turned into more complex demonstration devices. With the advent of reliable lighting technology "transparencies" or slides were introduced to astronomical education.[24] Models such as planetariums and slides were often used together. King writes about a lecture in 1824:

> Supplementing a large tellurium and a "New Transparent Orrery" are sets of transparencies of the sun, moon, planets, and nebulae painted by a M. D'Arcy "from original Drawings by Dr. Herschell, made from actual Observation with his Forty Feet Telescope, the largest and most perfect Instrument in the World." Visitors will see the planets "exactly as they appear in Dr. Herschell's Telescope with a power of Six Thousand Degrees, mountains and craters on a moon-map 7 feet in diameter, the bright comets of 1758 and 1811."[25]

What is important about this 1824 show is that it demonstrates that instruments, models, and images were used side-by-side and did not replace one another. Further, the show was centered around some of the most recent achievements, both technically and scientifically. Finally, it points at what will become later in the nineteenth century the most important educational medium, the lantern projector.

Lanterns as Instruments and Images

During the eighteenth century the lantern projector had become a common device for both study and demonstration all over Europe. It had gone through several changes but the principle had always stayed the same: a light source is placed in a wooden or metal box where the source's illuminative power is increased by means of a concave mirror and the light rays are made parallel by means of a condenser lens. A painted or photographic transparent slide can be placed in front of the condenser lens, where the light shining through this slide becomes enlarged by means of a focusable object lens and is projected onto a screen. In 1787 English optician and instrument maker G. Adams had replaced the faint candle and oil burner with the newly invented Argand burner. With this powerful light source the lantern projector could be used in large halls, and in 1798 the first public entertainment show took place in Paris. A few years later this so-called phantasmagoria entertainment had spread all over Europe, and workshops in London and Paris produced large quantities of such projectors. Though the construction principle remained unchanged the light source underwent several changes: by the mid-nineteenth century many households had been connected to town gas supplies, and lanterns with gas burners were introduced. The lime burner had been invented for use in lecture theaters, allowing the projection of images over a long distance. By the middle of the nineteenth century, lantern projectors found their way into households, public halls, and lecture theaters.[26]

In order to better understand the practice required to project historic lantern slides and to learn about the effect on the audience I decided to replicate some of the historic magic lantern demonstrations. It is difficult to find complete sets of nineteenth century astronomical slides as most of them are in the hands of either museums or collectors. A good example of such a collection is that of movie producer Werner Nekes. Nekes' collection is interesting especially because it contains some of the oldest mechanical astronomical slides known. Such slides project stellar constellations by means of fine, handmade perforations that let the light shine through and resemble the stars. These slides were produced, for example, by William Harris of London, around 1800.[27] By the middle of the nineteenth century astronomical slides were produced widely and instrument makers included such well-known names as Dollond or Negretti and Zambra (see figure 1.2). Standard representations of celestial movements by means of mechanical slides would include the rotation of the earth, comet movements around the sun, the sun and zodiac signs, the sun and the rotation of planets, the sun and the elongation of the inner planets, and the northern or southern sky and the Milky Way.[28] However, astronomical lantern slides in the middle of the nineteenth century were not limited to the representation of mechanical movements. Often they would include colorful and artistic images of celestial objects such as stellar constellations, the surface of the moon, Jupiter and its satellites, Saturn and its ring, comets with tails, or nebular objects.[29]

To get an impression of what was available in terms of astronomical slides, it is worthwhile to consult the trade catalogs of that time, for example, that of Negretti and Zambra of London.[30] The instrument trader had on offer thirty-eight astronomical slides of various astronomical illustrations. Further, the company offered ten movable lantern slides, listed as:

- The solar system
- The earth's annual motion around the sun
- Spring and neap tides and moon's phases
- Motion of Venus and Mercury
- A comet moving near the sun
- Rotundity of the earth
- Diurnal motion of earth
- Annual motion of the earth around the sun, with lunation of moon
- Eclipses of the sun and transit of Venus
- Eclipses of the moon, partial and total

One of most eminent contemporary collectors of magic lantern slides is Professor Wilhelm Wagenaar, Zeist, The Netherlands, who maintains a reconstructed nineteenth-century magic lantern theater and owns one of the most significant collections of both slides and lanterns. Professor Wagenaar was kind enough to give me access to his collection and helped me replicate and understand historic demonstration techniques. What is most striking when one looks at nineteenth-century astronomical slides such as found in Professor Wagenaar's collection are their colors. Unlike books of the same period, where illustrations were mainly black and white, lantern slides showed a vast range of colors. However, colors are not the only difference between book illustrations and slides at that time. Mechanical slides allowed demonstration of movements whereas book illustrations remained mainly static. Further, one should not forget that many of the generously illustrated books were not cheap and could be afforded only by a rather small group of the general population. This stood in contrast to the generally modest entrance fee charged for lantern shows or lectures.

Essential requirement for any demonstration or show was the lantern projector. I have already described its main features above: a box, a light source, a mirror, a condenser lens, a

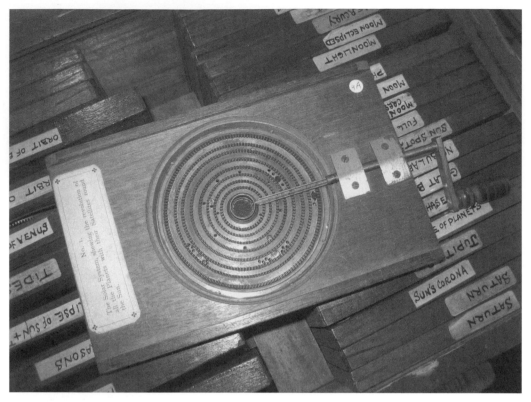

FIGURE 1.2
Mechanical slide for the demonstration of planetary motion, on box containing a complete set of astronomical slides, by Dollond, London, ca. 1850. Courtesy Klaus Staubermann.

slide holder, and an object lens (and a screen). Although the appearance of a projector could change significantly, the construction feature remained more or less the same.[31] Also, the materials during the nineteenth century remained more or less the same: glass for the optical components, brass or iron for the body and frame, and wood for the slide holder.[32] Only the light sources changed with the development of lighting technology during the nineteenth century: from candlelight to gas light, to oxy-hydrogen light, and finally to electric light. Oxy-hydrogen illumination for a long time remained the most common light source because of its illuminative power. However, it can also be considered to be the most difficult to control and accidents during the nineteenth century were not unusual. I do not want to go into any detail here regarding the control of the artificial light source and instead refer to what I have written about it elsewhere before.[33] Making light work required practice with the apparatus, the image, and finally the audience. Producing an image meant controlling its brightness and therefore the intensity of the colors of the projected image by means of flame, filters, and lenses. This was a skill projectionists had to develop.[34] Also, cleanness and order were an essential prerequisite for a successful show. Any dirt or dust on the slide, the lens, or the screen would become visible in the dark. Further, the projectionist had to move and work in the dark unhindered and had to rely on the arrangements made before the show, for example, the ordering of slides or sufficient lighting resources (see figure 1.3).[35]

Once the lantern had been set up, the light source been ignited, and the image focused, the show could begin. From the slides projected during my attempt to understand the making of a historic slide show it became apparent that simply showing images is not enough; rather, it can be dull and tiring for the audience. To engage the audience it was important to both prepare an interesting lecture and to choose from the slides at hand accordingly. Certainly, the stories told were as important as the images shown. Further, the lecture should be given without notes rather than read out. If the projectionist wanted to work with written notes, he had to take care that sufficient light was available at his workplace and at the same time did not blind the audience. In his choice of images the projectionist was not limited to astronomical slides. He could easily combine them with, for example, biblical or geographical motives and themes. From the posters advertising public lantern displayed in the nineteenth century it is evident that astronomy formed only one subject among many.

Contemporary projectionists were aware that the lecture was as important as the slides themselves. However, one would not succeed without the other. Lecture and slides had to be carefully synchronized, and the projectionist should not choose too many or too few slides. Also, the spectator was supposed to enjoy the images shown. "Good" slides would be shown longer, "bad" ones less long. Sitting in the dark would tire the audience or would make it indifferent, hence the lecture should not become too long or breaks should be inserted. Musical interludes could help the audience to relax during the course of a lecture. Finally, a test run of lecture and slide show was recommended to prevent any kind of mishap.[36] If the projectionist wanted to be on the safe side, he would purchase one of the combined printed lectures and slide sets available during the second half of the nineteenth century. More experienced demonstrators would consult popular astronomy books for information and motives.[37]

Producing a convincing slide was as difficult as producing a convincing show. A skill required by early projectionists was the painting and coloring of the images. Illuminating, enlarging, and projecting a slide image could change the appearance of an image on a screen and its effect on the audience significantly. The projectionist had also to take into account the fact that an image was rarely shown alone but was part of a sequence of images. Colors had to match within such a sequence and the effect of these colors on the spectators had to be considered carefully. Colors that appeared real in daylight could look unreal in lantern light: the artificial light and its different effect from daylight had to be taken into account. When the magic lantern came to be mass produced in the nineteenth century, awareness and skill for controlling artificial projections were developed for a wide audience.[38]

Lantern slides normally were made from glass and were produced in standardized formats. Several formats were in use and often depended on the country where they were produced or used. Also, the size increased with the development of projection technology during the nineteenth century. Glass slides, especially when commercially produced, would often be framed in wood to make them easier to handle and less prone to damage. Although slides of different sizes could be used with most of the lanterns (unless the slide became too large),

FIGURE 1.3
Hand of projectionist while moving the mechanical slide by means of a crank. Courtesy Klaus Staubermann.

it was not advised since the use of smaller slides together with larger ones would make apparent their lack of detail.[39]

Thomas Young in 1807 pointed out that it was essential for projecting slides that they "are made perfectly opaque, except where the figures are introduced, the glass being covered, in the light parts, with a more or less transparent tint, according to the effect required."[40] The production of slides required considerable skill. Glass plates of sufficient transparency had to be chosen and cleaned carefully. The outline of the images had to be drawn, first on paper and then copied with water color on glass plates. Then, the outlines were filled in with water color, starting with the lightest and finishing with the darkest. After every coloring it was dried and the surface varnished. If necessary, shadows were added and bright spots scratched out with a knife.[41]

I have already pointed out that the colors chosen for a series of slides had to be consistent in order to be convincing. Also, the appearance of colors can be quite different depending if the slide is held against a light source, such as a lit screen, or the light shines through a slide and is projected onto a screen. Colors had to be chosen that were as transparent as possible and could be dissolved in water completely. Water colors were favored to oil colors because in the latter, pigments were not dissolved completely and could make the image opaque. Colors that were not fast to light had to be avoided. Either gum Arabic or honey could be added to the color to protect the image from scratches or mechanical wear. Colors seen in nature could look very different when being projected on a screen and intuition had to be used to identify the correct tone.[42]

Once the images were painted they had to be prepared for later use. The surface was covered either by varnish or a second glass plate. The edges were covered by means of paper stripes or wooden (or metal) framing. All slides were marked with labels so they could be identified easily, normally on the thin side of the frame. The labels would also help to check the orientation of the slide and avoiding projecting the slide upside down. The slides were then stored in boxes made from wood or cardboard, often containing fifty or one hundred slides or even more, to protect them from light and humidity. An inventory or catalog would accompany the set of slides. Whenever the slides were shown they had to be checked before and after the show and cleaned or repaired if necessary.[43]

In order to produce convincing images the author had to familiarize himself with the subject. Motives such as the sun's corona or a comet's tail required both astronomical knowledge and skill in transforming the astronomical phenomenon into something visual. Imagination was needed but only within reasonable limits. Especially the poor contrast of astronomical objects, such as the stripes of Jupiter, posed a true challenge to the practitioner.[44] Mechanical slides had to be designed in such a way that, on the one hand, the proportions of all objects matched each other and, on the other hand, all objects were clearly visible. Still, many nineteenth-century slides seem to originate more from the visual imagination of their maker than from physical reality.

Reception and Audiences

What effect did the shows have on the audience? Surely, the audience was impressed by what was shown to them. A lantern show, if well-presented, could convey wonder and amazement in a way a book illustration could not. During the historic lantern shows restaged with the help of Professor Wagenaar, the audience responded positively and attentively. The colorful images stimulated fantasy and imagination—and often questions. Mechanical slides that demonstrated movements were always greeted with delight and received special attention. Almost all of the images were judged as truthful. Most surprisingly, the audience would come to a consensus on what they saw. For example, the lunar surface, although drawn based on artistic imagination, was experienced as convincing by most of the spectators. How can one explain such an impression of consensus of reception? The historian Wolfgang Schivelbusch in his book *Disenchanted Night* has argued:

> In light-based media, light does not simply illuminate existing scenes; it creates them. The power of artificial light to create its own reality is only revealed in darkness. The spectator sitting in the dark and looking at an illuminated image gives it his whole attention. The spectator in the dark is alone with himself and the illuminated image, because social connections cease to exist in the dark.[45]

It is worthwhile taking a closer look perhaps at how astronomical lantern performances were received in the nineteenth century. Three factors can be identified that helped to establish consensus among audiences: (1) a strong tradition of visual astronomical imagination among both astronomers and the public, (2) the ability of lantern show audiences to perceive the images as convincing, and (3) a somehow standardized mode of perception from the nineteenth century onward.

Gerald Holton has argued in his work *Scientific Imagination* that as early as in the seventeenth century astronomical studies had been translated into mental models with an explanatory visual component. He shows how Galileo's comparison of lunar surface structures with the Alps helped early observers to see mountains. By referring to a familiar analogy, visual consensus could be established more easily.[46] Holton shows further how this form of visual translation or analogy failed in other disciplines, such as quantum physics.

Terry Castle, in her work on nineteenth-century lantern shows, has asked how convincing such images were. She argues that "most contemporary observers stressed the convincing nature of a show."[47] It is difficult nowadays to conceive that the often crudely designed images could have had a convincing effect on the spectator, but we must accept that indeed they had. Also, one should not underestimate the spectators' will "to believe" in what they saw.

The third feature specific for spectators' modes of perception in the nineteenth century was standardization. Media historian Jonathan Crary has convincingly argued that in the nineteenth century, spectators' perception and imagination became calculable, regularizable, measurable, and exchangeable. Crary has shown that the spectator's visual imagery cannot be understood without the apparatus that produced it and vice versa.[48] Both the spectator and the projection apparatus became standardized in the nineteenth century. One outcome of this process of standardization was spectators' increasing ability to come to a consensus of what they saw at a lantern show.

That the wonder and astonishment of astronomical slides could cross almost all social boundaries is demonstrated by an occurrence reported by Friedrich Simon Archenhold, then director of Berlin's public observatory. On July 4, 1898, he received a message from Germany's Dowager Empress Victoria that she would like to observe the moon through the observatory's giant telescope the next day. However, because of poor weather, observing had to be cancelled and Archenhold was ordered to the imperial castle the same day instead.

FIGURE I .4
Slide projection of the imagined surface of the Moon, ca. 1850. Courtesy Klaus Staubermann.

Archenhold describes how he first demonstrated some astronomical models and then some slides he brought with him to the empress. Reportedly, the empress was especially impressed by images of the lunar surface and when Archenhold demonstrated the tidal effect of the moon by means of a mechanical lantern slide she exclaimed; "Pity, that during my youth, education was not yet supported by such beautiful means of demonstration."[49] The story continues that the empress was so impressed that one week later she attended one of Archenhold's public lantern lectures, unannounced and in the company of two equally impressed ladies-in-waiting. Archenhold's example nicely shows that not only could astronomical lantern shows cross almost all social boundaries, but also that during the nineteenth century demonstration instruments had not disappeared because of images, but they had, in fact, helped to produce them.

Notes

1. Newcomb and Engelmann, *Populaere Astronomie*.
2. Meyer, *Illustrierter Leitfaden der Astronomie*, 1–96.
3. Anonymous, *An Illustrated Catalogue*.

4. Hess, *Himmels- und Naturerscheinungen*, 1–21.

5. Hentschel, "Drawing, Engraving, Photographing, Plotting, Printing."

6. Gerhardt, *Geschichte der Druckverfahren*.

7. Griffiths, *Prints and Printmaking*.

8. Utzt, *Die Bilder der populären Astronomie*. The numerous technical, cultural, and social factors causing a rise of demand for images have just begun to be studied. See also Duerbeck, "Book Review."

9. Utzt, *Die Bilder der populaeren Astronomie*, 38.

10. John, *August Heinrich Christian Gelpke*.

11. Ibid., 106.

12. Elly Dekker, *Globes at Greenwich*, 566.

13. Zinner, *Deutsche und Niederländische Astronomische Instrument*, 167–77. See also Bud and Warner, *Instruments of Science*, 28–31, and Turner, *Early Scientific Instrument*, 16–18.

14. For some fine examples of combined demonstration instruments see Dolz, *Erd- und Himmelsgloben*, 89–97.

15. I am grateful to Michael Korey for various stimulating conversations here.

16. Zinner, *Deutsche und Niederlaendische Astronomische*, 40–47. For examples, see Dekker, *Globes at Greenwich*, 3–176.

17. For the large variety of nineteenth-century demonstration devices produced see, for example, Stephenson, Bolt, and Friedman, *The Universe Unveiled*.

18. Haase, *Alte Karten und Globe*, 61.

19. Warner, "Celestial Technology."

20. Daumas, *Scientific Instruments*, 12–13.

21. Putsche, *Planetarium*, i–viii.

22. John, *August Heinrich Christian Gelpke*, 104.

23. Beech, "The Cometarium by John Taylor."

24. King, *Geared to the Stars*, 309–68.

25. Ibid., 317.

26. Staubermann, "Making Stars," 440–41. See also Paul Liesegang, *Zahlen und Quellen*.

27. von Dewitz and Nekes, *Ich sehe was*, 162–63.

28. Ibid., 163.

29. For examples of mid-nineteenth-century astronomical slides see www.olemiss.edu/depts/u_museum/Millington/slides2.htm (accessed April 2009).

30. Negretti and Zambra, *The Magic Lantern Dissolving Views*, 8.

31. Pepper, *The Boys' Playbook of Science*, 346–55.

32. Wright, *Optical Projection*.

33. Staubermannn, "The Trouble with the Instrument."

34. Young, *A Course of Lectures*, 426

35. Hauberrisser, *Anleitung zum Projizieren*, 110–13.

36. Lettner, *Skioptikon*, 64–66.

37. Liesegang, *Die Projektions-Kunst*, 302–3.

38. Schiendl, *Die optische Laterne*, 124.

39. A format often employed for public lectures was, for example, height 120 mm, width 180 mm, and depth 10 mm.

40. Young, *A Course of Lectures*, 426–27.

41. Hrabalek, *Laterna Magica*, 28–31.
42. Schiendl, *Die optische Laterne*, 124–33.
43. Hassack and Rosenberg, *Die Projektionsapparate*, 129–86.
44. Valier, *Das Astronomische Zeichnen*.
45. Schivelbusch, *Disenchanted Night*, 220.
46. Holton, "On the Art of Scientific Imagination."
47. Castle, "Phantasmagoria."
48. Crary, *Techniques of the Observer*, 16–17.
49. Archenhold, "Kaiserin Friedrich."

References

Anonymous. *An Illustrated Catalogue of Astronomical Photographs*. Chicago: Chicago University Press, 1931.

Archenhold, Friedrich Simon. "Kaiserin Friedrich, eine Freundin der Astronomie." *Das Weltall* 1 (January 1900): 189–91.

Beech, Martin. "The Cometarium by John Taylor." *Bulletin of the Scientific Instrument Society* 88 (2006): 28–32.

Bud, Robert, und Deborah Jean Warner *Instruments of Science: An Historical Encyclopedia*. London, New York: Science Museum and National Museum of American History, Smithsonian Institution, in association with Garland Publications, 1998.

Castle, Terry. "Phantasmagoria: Spectral Technology and the Metaphorics of Modern Reverie." *Critical Inquiry* 15 (1988): 26–61.

Crary, Jonathan. *Techniques of the Observer: On Vision and Modernity in the Nineteenth Century*. Cambridge, MA: MIT Press, 1990.

Daumas, Maurice. *Scientific Instruments of the Seventeenth and Eighteenth Centuries and Their Makers*. Translated and edited by Mary Holbrook. London: Batsford, 1972.

Dekker, Elly. "The Copernican Globe: A Delayed Conception." *Annals of Science*, 53 (1996), 541–66.

Dekker, Elly. *Globes at Greenwich: A Catalogue of the Globes and Armillary Spheres in the National Maritime Museum, Greenwich*. Oxford and New York: Oxford University Press and the National Maritime Museum, 1999.

Dolz, Wolfram. *Erd- und Himmelsgloben*. Dresden: n.d.

Duerbeck, Hilmar. "Book Review—Astronomie und Anschaulichkeit." *Journal for Astronomical Data*, 10 (2005): n.p.

Gerhardt, Claus W. *Geschichte der Druckverfahren*. Stuttgart: Hiersemann, 1975.

Griffiths, Antony. *Prints and Printmaking*. Berkeley: University of California Press, 1996.

Haase, Yorck Alexander. *Alte Karten und Globen in der Herzog August Bibliothek Wolfenbüttel*. Wolfenbuettel: Herzog August Bibliothek, 1972.

Hassack, Karl, and Karl Rosenberg. *Die Projektionsapparate*. Wien and Leipzig: 1907.

Hauberrisser, Georg. *Anleitung zum Projizieren*. München: 1912.

Hentschel, Klaus. "Drawing, Engraving, Photographing, Plotting, Printing: Historical Studies of Visual Representations, esp. in Astronomy." In Klaus Hentschel and Axel Wittmann, *The Role of Visual Representation in Astronomy*. Frankfurt: Harri Deutsch, 2000, 11–43.

Hess, Wilhelm. *Himmels- und Naturerscheinungen in Eindruckblättern*. Leipzig, 1911.

Holton, Gerald. "On the Art of Scientific Imagination." *Daedalus*, 2 (1996): 182–208.

Hrabalek, Ernst. *Laterna Magica*. München: Keyser, 1985.

John, Klaus. *August Heinrich Christian Gelpke—ein Astronom am Collegium Carolinum zu Braunschweig*. PhD dissertation, Braunschweig University, 2004.

King, Henry C., in collaboration with John R. Millburn. *Geared to the Stars*. Toronto: University of Toronto Press, 1978.

Lettner, G. *Skioptikon*. Leipzig, 1905.

Liesegang, Paul. *Die Projektions-Kunst*. Leipzig, 1909.

Liesegang, Paul. *Zahlen und Quellen zur Geschichte der Projektionskunst und Kinematographie*. Berlin, 1926.

Meyer, M. Wilhelm. *Illustrierter Leitfaden der Astronomie, Physik und Mikroskopie*. Berlin: 1892.

Negretti and Zambra. *The Magic Lantern Dissolving Views*. Catalog 8. London: n.d.

Newcomb and Engelmann. *Populäre Astronomie*. Leipzig, 1922.

Pepper, John Henry. *The Boys' Playbook of Science*. London, 1860.

Putsche, C. W. E. *Planetarium*. Weimar, 1805.

Schiendl, V. *Die optische Laterne*. Karlsruhe, 1896.

Schivelbusch, Wolfgang. *Disenchanted Night: The Industrialisation of Light in the Nineteenth Century*. Translated by Angela Davies. Oxford and New York: Berg, 1988.

Staubermannn, Klaus. "The Trouble with the Instrument: Zoellner's Photometer." *Journal for the History of Astronomy* 31 (2000): 323–38.

Staubermann, Klaus. "Making Stars: Projection Culture in Nineteenth Century German Astronomy." *British Journal for the History of Science* 34 (2001): 439–51.

Stephenson, Bruce, Marvin Bolt, and Anna Felicity Friedman. *The Universe Unveiled*. Cambridge and New York: Cambridge University Press; Chicago: Adler Planetarium & Astronomy Museum, 2000.

Turner, Anthony. *Early Scientific Instruments Europe, 1400–1800*. New York: Philip Wilson Publishers for Sotheby's Publications, 1987.

University of Mississippi "More Magic Lantern Slides," www.olemiss.edu/depts/u_museum/Millington/slides2.htm (accessed April 2009).

Utzt, Susanne. *Die Bilder der populären Astronomie des 19. Jahrhunderts*. Frankfurt: Harri Deutsch, 2004.

Valier, Max. *Das Astronomische Zeichnen*. München: Verlag Natur and Kultur, 1915.

von Dewitz, Bodo, and Werner Nekes. *Ich sehe was, was Du nicht siehst! Sehmaschinen und Bilderwelten*. Göttingen: Steidl, 2002.

Warner, Deborah. "Celestial Technology." *Smithsonian Journal of History* 2, no. 3 (1967): 35–48.

Wright, Lewis. *Optical Projection*. London, 1891.

Young, Thomas. *A Course of Lectures on Natural Philosophy and the Mechanical Arts*. London, 1807.

Zinner, Ernst. *Deutsche und Niederländische Astronomische Instrumente des 11.–18. Jahrhunderts*. Muenchen: Beck, 1956.

CHAPTER 2

The Enlightened Microscope

WORKING WITH EIGHTEENTH-CENTURY SOLAR MICROSCOPES

Peter Heering

Introduction

Solar microscopes are devices that were developed in the 1740s and became extremely popular throughout the second half of the eighteenth century. However, in the nineteenth century, their popularity declined and they no longer played a role in natural philosophy. Nowadays these instruments are found in showcases of most science museums, and in most cases only one or two of several instruments in the collection are on display.

But what is a solar microscope, and what did eighteenth-century people do with this device? This chapter focuses on an attempt to reenact the experiments that were carried out with solar microscopes. In the first part of the chapter, the instrument will be discussed from a historical point of view. In consulting textbooks of the eighteenth and nineteenth centuries I will show what kind of notions about the solar microscope had been developed. In the second part, I will analyze two instruments from the collection of the Deutsches Museum from a technological point of view. In doing so, I am relying on experiences made in working with these devices. In the third part, I will discuss these experiments and try to develop an understanding of the results that could have been achieved with these devices. Finally, I am going to draw some conclusions on the role of solar microscopes and what can be learned in using them.

Looking into Books

In a first approach to discuss what a solar microscope is and what can be done with it, various textbooks can be consulted: By looking at one from the eighteenth century the following could be found: "The ordinary solar microscope for transparent objects consists of a tube, a plain mirror, an illuminating lens and an ordinary Wilson microscope."[1] Apart from this short sketch of the major parts of the instrument, a description of the solar microscope's working principle is given, sometimes together with an illustration (see figure 2.1). The one reproduced here is particular as it does not only show the instrument, but also major aspects of the working principle: Sunlight is reflected by a plain mirror onto a condensing lens. The resulting light cone is focused on the (transparent) microscopic object. The image of this object is projected by a microscope, which was in most cases a single lens Wilson pocket microscope.

The solar microscope is used in a darkened room (see figure 2.2), the mirror being outside of the shutter of the window. The images were either projected onto a large screen in

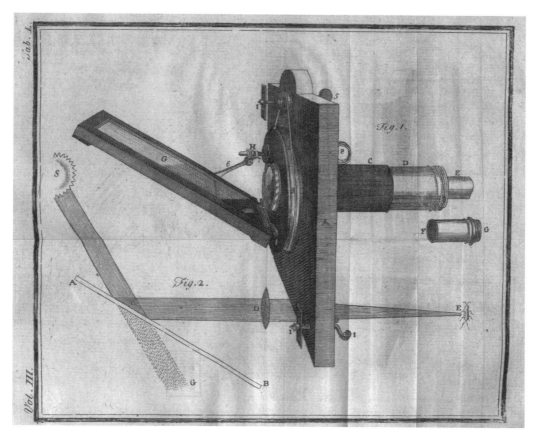

FIGURE 2.1
Solar microscope. From Benjamin Martin's *Philosophia Britannica: Neuer und faßlicher Lehrbegrif der Newtonschen Weltweisheit, Astronomie und Geographie* (1772). Courtesy Deutsches Museum, Munich.

order to be shown to an audience—the plate showing four chairs on which the viewers can be seated. However, the solar microscope can also be used with a smaller screen in order to make microscopic drawings; due to the intensity of the light it should be possible to draw the image on the reverse side of the paper. This means that the drawing hand did not produce a shadow on the image, thus making the drawing less complicated.

Several authors pointed out the high quality of the projected images thus produced. In this respect it is not surprising that some of them were instrument makers who obviously wished to promote their products. However, they include authors who cannot be alleged to have a commercial motivation to praise the quality of the images: "The sunbeams, directed from the illuminating lens onto the transparent object create on the opposite wall . . . a clear and beautiful image that much enlarged than nobody could imagine who had not seen it himself."[2]

However, examining nineteenth-century publications with respect to solar microscopes gives a completely different picture. A very characteristic and influential rating of the images produced by solar microscopes was published by British physician and microscopist Charles R. Goring in 1827: "The image of a common solar microscope may be considered as a mere shadow, fit only to amuse women and children. . . . The utmost it can do is to give us the shadow of a flea, or a louse as big as a goose or a jackass."[3]

These differences are not only limited to the accounts of the quality of the images produced by a solar microscope, but also the status of the instrument changes significantly when textbooks from the eighteenth century are compared with those from the nineteenth. A typical example for an eighteenth-century publication reads as follows: "The solar microscope is an extraordinary and very interesting instrument. It is very capable to expand the progress in natural philosophy and physics by the facility to see enlarged and without any eyestrain and by several persons at the same time extremely small objects."[4] A completely different image

FIGURE 2.2
Solar microscope in situ. From M. F. Ledermüller's *Nachlese seiner mikroskopischen Gemüths- und Augen-Ergötzung* (1762). Courtesy Deutsches Museum, Munich.

was produced in nineteenth-century publications: "We are also remarking that the solar microscope is not useful for scientific investigations and serves more for entertainment."[5]

Looking at Instruments

Although one gets a fairly good idea from studying textbooks on what could have been done with a solar microscope, some questions remain open. Particularly with respect to technical details of the instruments as well as to the possible quality of the projections, more research seems to be justified. To do this, it seems to be useful to go into the stores of the museum and examine the items that are kept under the label "solar microscope": Actually in many cases there is not only one solar microscope but several; for example, in the collection of the Deutsches Museum fourteen instruments can be found, eleven at the Universiteitsmuseum Utrecht, sixteen at the Museum for the History of Science Oxford, and forty-two at the Science Museum London.[6] One thing that becomes very obvious immediately when one starts to examine the instruments is that there is no such thing such as "the solar microscope" but a variety of devices that are all based on the working principle sketched above. However the technical realization of the device differs significantly. To illustrate this I am going to discuss two of the solar microscopes kept at the Deutsches Museum in more detail.

The first of these devices is actually not to be found in the store but in a showcase in the optics gallery (figure 2.3). It is signed by John Dollond and was purchased by the museum in 1915 from a local instrument dealer. The entire instrument is made of brass, together with the instrument goes a wooden case that contains some additional lenses and ivory sliders in which preparations are kept. As Dollond was one of the major London instrument makers

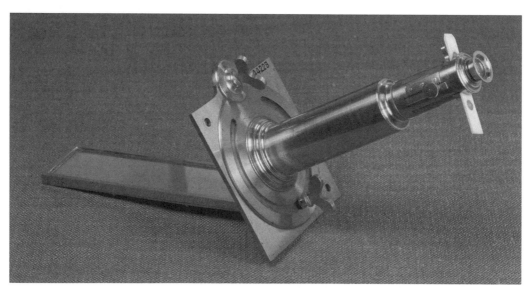

FIGURE 2.3
Solar microscope, John Dollond. Courtesy Deutsches Museum, Munich.

and particularly well-known for his optical instruments, it can be assumed that this device is a state-of-the-art example of the last quarter of the eighteenth century.[7]

The second instrument is a solar microscope made by Friedrich August Junker in the early 1790s. Actually there is no complete solar microscope of Junker but one instrument where the mirror, the mechanism for its adjustment, and the condensing lens are missing, another where the tube, the microscope, and the (wooden) sliders with the preparations are lost. Taking both instruments together one gets a complete solar microscope that can be used (figure 2.4).[8] Compared to Dollond, Junker's background is entirely different. He was a field chaplain and decided to produce solar microscopes particularly for school purposes. In doing so, he attempted to make his instruments as cheap as possible; consequently his solar microscope costs about 10 percent of the price of an English one, such as Dollond's.[9]

While working with the instrument, three operations have to be executed: the slider has to be moved in the microscope, the image has to be focused, and—the most frequent thing to be done—the mirror has to be readjusted as the seemingly motion of the sun has to be compensated. The latter activity has to be carried out every four to five minutes, as had already been observed by Ledermüller in his 1762 description: "But what is more arduous than to draw from the solar microscope as the sun remains less than four minutes in its place but permanently moves on. . . . It thus requires a very proficient and swift hand to sketch an object with the solar microscope."[10]

The technical solution of the problem on how to compensate for the movement of the sun differs significantly between the two instruments. In case of Junker's instrument the mirror is fixed to a wooden disk by a brass needle serving as an axis. At the frame of the mirror a string is fixed, its other end is attached to a conical wooden pivot that is inserted into a hole in the disk. By turning this pivot the string is shortened or lengthened, consequently the mirror is lowered or raised. The entire disk together with the mirror can be turned with the help of a

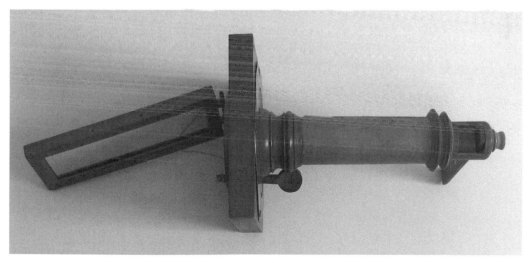

FIGURE 2.4
Solar microscope, Friedrich August Junker. Courtesy Peter Heering.

bushing that is placed on the inside and serves also as the holder of the tube. As a result, whenever the disk is turned, the tube and the microscope together with the slider are also rotating.

This was probably one of the reasons why Dollond chose another technical realization for the mirror adjustment. Remarkably, most English solar microscopes from the last quarter of the eighteenth century—and also several made on the continent—have a very similar mechanism. The mirror is again fixed on what appears on the first look to be a brass disc. The angle between the mirror and this disc can be changed with an endless screw and a sector, thus being able to move the mirror toward or away from the disc. On the other side, the disc is covered with a brass plate with a slot in which the endless screw can be moved. At the upper part is another screw—if this is turned, the brass disc will also rotate. Actually, it appears to be not a simple disc but a toothed locked washer; the screw above it is connected to a toothed wheel. This construction enables a very precise adjustment of the mirror; however, from this construction results also another advantage. The tube and the microscope are no longer mechanical attached to the disc with the mirror but to the plate that covers it. Consequently, neither the microscopic lens nor the specimen is moving during the readjustment of the mirror; as a result, the image does not turn on the screen when the mirror is readjusted.

In a similar manner the mechanism of focusing the image on the screen differs significantly; again, the solution found in Dollond's instrument is also used in most British and many continental solar microscopes from the last quarter of the eighteenth century. The lens can be moved away or toward the slider with the help of a rack and pinion mechanism. This makes possible a very sensitive and accurate adjustment of the lens.

In Junker's instrument the image on the screen can also be focused by changing the distance between the lens of the microscope, and the object. The technical solution, however, is completely different: In his solar microscope, the lens is inserted in a wooden holder with a screw thread of several millimeters. This holder is screwed into a wooden cylinder; however, due to the length of the screw thread it is not necessary to tighten the holder. Therefore, it is possible to screw the holder either deeper into the cylinder or out of it, thus changing the distance between the lens and the specimen.

Both instruments were sold together with sliders that contained ready-made preparations. Dollond's sliders are made of ivory, Junker's are wooden. All sliders have circular apertures (with a shoulder at one end) that contain the specimens. These were placed between two discs (either mica or glass), the upper one being fixed with a brass circlip. In case of Dollond's instrument, six consecutively numbered sliders exist; each of them contains four specimens. In addition, another two sliders—one of them broken—are found in the box. Also in the box is a small ivory capsule that contains spare mica discs and circlips. Together with Junker's instrument, five sliders were sold, each of them containing five specimens.

Seeing with Instruments

The room has to be darkened completely to use the instrument. As Henry Baker points out: "When this Microscope is employed, the room must be rendered as dark as possible: for on the Darkness of the Room, and the Brightness of the Sunshine, depend the Sharpness and

Perfection of your Image."[11] The instrument is placed into the shutter of the window (a wooden board in which a metal plate is inserted). In doing so, the microscope is removed, thus it is possible to see the mirror and adjust it in a manner that the sun is reflected into the room. The adjustment is appropriate when a circular light spot appears on the screen. Then the microscope is put in place and the instrument is ready for use.

Even when the instrument is not equipped with a slider it becomes clear that the optical properties of the Dollond solar microscope are significantly better than the ones of Junker's. The image on the screen is a clear, almost constantly illuminated light circle, while the Junker solar microscope produces a circle with an irregular light intensity and a very streaky appearance (see figure 2.5). But it is not only the optical properties of the Dollond instrument that make it appear superior to the Junker. When the necessary procedures to adjust the mirror and to focus the image on the screen are compared in a darkened room, it becomes obvious that also the mechanism of the Dollond instrument is superior to that of Junker's solar microscope. The image on the screen produced with Dollond's instrument can be focused in a very precise manner. Moreover, the image remains very clear on the screen. This is different from working with the Junker solar microscope: The adjustment of the holder of the lens is less accurate, it is necessary to change the direction of turning several times in order to find the best position. Furthermore, the image remains not on the same position on the screen, but seems to be gliding over it or even partly leaves the screen.

The optical superiority of Dollond's solar microscope has, of course, consequences for the images that can be projected. They are very clear and produce for modern viewers the impression of a slide show. The image itself is homogeneously illuminated, the boundary is very clear as it is formed by the aperture in the slider. But it is not only the clearness of the boundary that gives the impression of a high quality and very stable representation of the object. The image itself is crucial in this respect as it appears to be also extremely well focused when looked at from a distance. Yet, in several cases this impression turned out to be misleading as a close-by inspection revealed that these images were not perfectly focused.

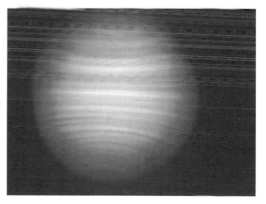

FIGURE 2.5
Junker and Dollond solar microscope images. Light circles produced with the Dollond instrument (left) and the Junker instrument (right). Both instruments were used without the microscope part. Courtesy Peter Heering.

However, with some experience (or a person standing close to the screen and giving a feed-back during the process of focusing) it is possible to produce images on the screen that are actually focused. [12]

What is striking with respect to all the projections is that the preparations are soiled. As the dissected objects had been placed between two coverslips that were fixed with a circlip,[13] it was only possible to clean the outside of the coverslips. As it turned out, that improved the quality of the image but there was still some dirt between the coverslips that could not be removed.

The objects themselves are not identifiable unless one has already some experience in working with microscopic preparations. Like these two instruments, most solar microscopes were originally sold together with five or six readily prepared sliders—as it had been charac-terized for microscopes in general by Ian Hacking:

> The microscope became a toy for English ladies and gentlemen. The toy would consist of a microscope and a box of mounted specimens from the plant and animal kingdom. Note that the box of mounted slides might well cost more than the purchase of the microscope itself. . . . All but the most expert would require a ready mounted slide to see anything.[14]

From my understanding, Hacking gives here a simplified description of the status of microscopes in the eighteenth century: Although complaints can be found in publications from that period with respect to the nonscientific attitude of several users, the characteriza-tion as a toy is—as I will discuss later—an oversimplification that is probably influenced by the retrospective accounts from the nineteenth century.

However, there is another aspect of the sliders accompanying a microscope in which Hacking's characterization is oversimplifying: First of all, many sets did include, apart from the already mounted sliders, some that were still empty. Moreover, as already mentioned in the description of the Dollond solar microscope, this set also contained mica discs and brass circlips. In other words, the instrument maker provided everything necessary to make your own preparations. This was made explicit in Nairne's leaflet accompanying his instrument: "In the Partition L are 8 Ivory Sliders, in six of which are different Objects, a List of which are undermentioned; the other two, left vacant, are to be filled at Pleasure. . . . V is an Ivory Box, containing Talc at one End, and Wires at the other, fitted to the Holes in the Ivory Slid-ers, in Order to supply what may be lost, or to be changed at Pleasure."[15]

This set of sliders was accompanied by a printed leaflet that contained hints on how to use the solar microscope as well as a list of the objects in the sliders. Therefore, it was possi-ble for the initial users to develop an understanding of what they actually were seeing by using the sliders together with the leaflet. As I lacked this leaflet, I had to find other ways of developing an understanding of what actually can be seen. In this respect it was very help-ful that the list of samples accompanying Junker's instrument still existed. Several of the preparations that belonged to Junker's instrument are very similar to those belonging to Dol-lond's. It does not seem to be implausible that several specimens were chosen commonly for ready-mounted sliders, either because they were well-suited for projections or because they had a specific meaning in the context of microscopy.

With the help of Junker's microscope it became possible to identify about half of the preparations; the others could be identified with the help of colleagues who had a microscopic background. Thus it is possible to categorize the preparations in the sliders; the first slider contains plant parts (figure 2.6, top left), the second various fish scales (figure 2.6, top right), the third insect wings and a leg (figure 2.6, bottom left), the fourth part of a butterfly wing and plant semen, the fifth mostly destroyed preparations, and the sixth shavings of wood (figure 2.6, bottom right). As indicated already by this listing, several of the preparations have decomposed. It is very likely that these were specimens that are missing in comparison to any given series of preparations—such as a flea, a louse, or some mites. The objects in the sliders are grouped in a thematic order: insects' parts together in one slider, sectioning from wood together in another. Remarkably, the intact specimens that were together in

FIGURE 2.6

Dollond solar microscope slides. Re-enacted projection of original Dollond preparations using the Dollond instrument. All light circles had a diameter of about 1.60 meters; exposure time was 8 seconds. The preparations were a part of a plant (top left), a fish scale (top right), a segment from an insect's wing (bottom left), and a section of wood (bottom right). Courtesy Peter Heering.

the slides with the corrupted ones were animal. Thus, it seems plausible to suppose that the ones that were destroyed were also of animal origin and could have been objects such as the flea.[16]

Remarkably, none of the preparations that belong to Junker's instrument have disintegrated. Moreover, not only all the specimens have survived but also the list that identifies them. Again, the arrangement of the preparation in the sliders is based on content criteria.[17] However, although none of the specimens had decomposed, they were even more soiled than the ones of the Dollond instrument. There may be two reasons for this. The instrument was designed to be used in schools, and actually the sliders belong to the instrument that was donated to the museum by a school. It is very likely that the instrument had been used under school conditions and was thus not kept in a manner that reduced the contamination of the specimens. Moreover, inaccurate handling and resulting mechanical stress may have caused minute particles of the wood of the slider to break up and to be incorporated into the preparation.

However, the impression of the image on the screen is not dominated by these effects but by the optical ones. The images differ significantly from those that can be produced with the Dollond instrument.[18] One of the striking aspects with respect to the general impression is that the aperture of the slider is not completely illuminated. Therefore, the image is no longer clearly bounded but seems to vanish at the borders. Moreover, the illumination is fairly irregular; one area of the preparation appears brightly illuminated, but still not homogeneous. Other parts are poorly illuminated; still others are not to be seen on the screen. As a consequence, it is necessary to move the slider slightly up and down during the projection. Consequently, the image is changing during the presentation and the impression is generated that the object is "scanned."

Nevertheless, despite these drawbacks it is possible to look at the objects and—with the help of the list—to identify them without many problems: The first slider contains shavings of wood (see figure 2.7, top left), the second parts of various plants (figure 2.7, top right), the third parts of flying insects (figure 2.7, bottom left), the fourth eyes of an insect and of a crustacean, a louse, and a flea (figure 2.8, top), and the fifth animal particles such as a fish scale (figure 2.7, bottom right) or parts of the feather of a goose.

Comparing the images in figures 2.6 and 2.7, striking differences are evident. However, one has to be cautious as the pictures are misleading in several respects: First of all, the images projected with Junker's microscope do not appear stable but—owing to the optical properties of the instrument—are changing slightly. Moreover, the slider has to be moved up and down during the demonstration, thus the information gathered from the projection is more than it can be represented in one picture. Furthermore, the images are not as black and white as they seem to be in the figures. Although the color does not appear as one might expect, most of it seems artificial and not part of the adequate image of the object. Particularly in the projections with Junker's solar microscope, aberration causes significant occurrence of colors. The projections with Dollond's solar microscope also suffer from aberration effects; however, these are much weaker and become apparent only when standing close to the screen.

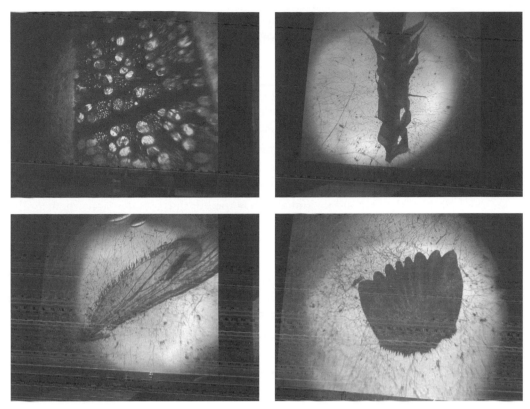

FIGURE 2.7
Junker solar microscope slides. Re-created projection of original Junker preparations with the Junker instrument The screen had a width of 1.60 meters; exposure time was 8 seconds. The preparations were a section of hazel wood (top left), a piece of moss (top right), a wing of a mosquito (bottom left), and a scale from a perch (bottom right). Courtesy Peter Heering.

Furthermore, the Dollond instrument also produced another color appearance. About every four minutes, in the upper right part of the image appears a golden coloring. This results from the mirror being no longer properly adjusted; the color impression is a result of some light being reflected from the brass frame. If nothing is done, this impression becomes stronger and finally starts to influence the image itself. However, an experienced demonstrator can take its initial appearance as a signal that the mirror needs some readjustment, using this indication makes it possible to do this without the viewer realizing this activity. This is different when working with Junker's solar microscope; due to the inhomogeneous appearance of the image and the wooden frame of the mirror no such indication for the necessity of the mirror adjustment could be identified. Consequently the mirror was either readjusted some five minutes after the last adjustment or when the image changed significantly. Moreover, due to optical difficulties and the mechanically less sensitive mechanism of the mirror regulation, it is not possible to readjust the mirror without the audience noticing.

One of the aspects that is striking with respect to the image and where the pictures are again misleading is the color of the image itself. Most of the specimens are colorless; initially it was unclear whether this has to be taken as a result of the intense illumination with focused sunlight or some other sort of ageing process on the one hand,[19] or whether projections with solar microscopes are unable to produce colored images. The latter could not be excluded; solar microscopes are also unable to make projections of opaque specimens, a limitation that is mentioned frequently in eighteenth-century literature. Moreover, if the criticisms of the nineteenth century are recalled, this raises the question about the possible quality of projected images. On the other hand, it is obvious that the specimens may have changed over the centuries, and not only because of the soiling of the specimens. Some of the specimens that belong to the Dollond instrument decomposed, thus it is not clear whether others might be changed to a lesser degree.[20]

Creating New Projections

In order to develop an understanding of what kind of projections may have been possible in the eighteenth century, I attempted to prepare new sliders. The importance of these sliders tended to be underestimated even in the historical situation: "The sliders do not seem to have the relevance so that they are considered as an important part of the microscope. . . . However, I am going to show that a good composition of them is very convenient."[21] Unfortunately, even Wiedeburg did not make explicit how to prepare the sliders. Thus it remained unclear how the specimens may have been treated in preparing a slider. However, Henry Baker devoted one chapter of his book to "preparing and applying Objects."[22] From this description it becomes clear that many specimens were simply placed between two glass discs that were fixed with the circlip.

In order to get an idea of what a solar microscope may have been able to show in the eighteenth century, a technician from Munich University was asked to prepare a slider with some objects that were not bleached or dyed.[23] This means of course that this would be a high-quality specimen that might even exceed the quality of historical ones. On the other hand, the organization of the large London workshops in the second half of the eighteenth century changed significantly and "heralds the era of industrial manufacture, characterized by the division of labour and serial production."[24] Consequently, within such an organization of the production of microscopic sets, it is very likely that sliders were produced in large numbers by specialists. Thus, it is very plausible that the specimens fulfilled high-quality standards, particularly those that were sold with a Dollond instrument. As a consequence, it seemed justified to use high-quality preparations that were made on the basis of techniques that were at least available in the late eighteenth century.

One might assume that things may be different with Junker's sliders as it is very likely that he prepared them himself: He made clear in his leaflet that the sliders would be wrapped in a piece of paper on which he had identified the specimens. The paper accompanying the Munich sliders is handwritten, although several of the specimens are described already in the leaflet. On the other hand, some of the specimens described in the leaflet are not to be found on this list. Thus, it is plausible that Junker picked specimens that were available to him. However, this does not mean that these were poor quality. On the contrary, in a short note announcing Junker's solar microscope, the author claims that he "has tested this solar microscope . . . the objects are very well chosen and nicely dissected."[25]

Apart from the preparations made by the technician, another attempt was made to prepare "new" sliders: as already mentioned, the Dollond instrument did not only contain six consecutively numbered sliders but two others. These contain two or even three preparations in each aperture of the slider; this seems to be fairly unusual for sold preparations. Moreover, as the set contains some circlips and mica discs, that is, all the materials necessary to make preparations with an empty slider, it is very likely that these sliders were actually prepared by one of the initial users. As the leaflet with instructions did not contain any hint of how to make preparations, it does not seem to be unlikely that the specimens were simply placed between the mica discs, particularly as this seems to correspond to Baker's description mentioned above. In a like manner I also attempted to make my own preparations.

The results of both attempts were very convincing, this becomes clear when the three fleas are compared (figure 2.8 shows the flea of the Junker instrument [top left], the one prepared by the technician [top right], and the one I prepared myself [bottom left], all projections were made with Dollond's solar microscope). The first thing that is striking lies in the poor quality of the old flea. Taking this as a standard, it can justly be argued—using Goring's expression— that the image is a mere shadow. The impression is dominated by the contamination of the preparation, but the flea itself is not very clear and the details cannot be made out. Thus it is understandable that such an image cannot be taken as being an indication of the quality of the instrument or, to be more accurate, this image seems to indicate the poor quality of the instrument. However, even if we just consider the third flea, which is simply placed between the two glass discs, it becomes obvious that the problem is more the preparation than the instrument: The flea can clearly be identified, although the body is still nothing but a shadow. However, details of the shape as well as the legs can be clearly recognized. Moreover, parts of the legs appear to be slightly transparent (which is hardly seen on the picture as this could only be reproduced black and white). These problems in reproducing the picture are even more relevant in the case of the flea prepared by a technician at Munich University, Heidrun Schöl. As it can be at least anticipated on the picture, the image of this flea is fairly transparent, even the body of the insect is colorful; the entire image of the insect appears yellow and coppery. The image of this preparation shows clearly that problems in producing images with the solar microscope are not necessarily due to deficits of the instrument, on the contrary. This becomes even more obvious when the image produced with Dollond's solar microscope is compared with the image of the same object projected with Junker's instrument. If the flea is taken as an example, it is striking that it appears also transparent and as colorful as in the projection with Dollond's instrument. The appearance of the image is still different as there are no clear borders to the image, but the object itself is small enough so that it is not affected by this deficit of the instrument.

At this point it should be questioned why the flea prepared by Schöl was that much more transparent than mine. It turned out that she had treated the flea with alcohol and xylene in

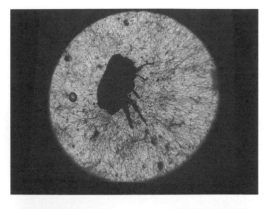

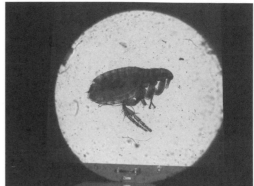

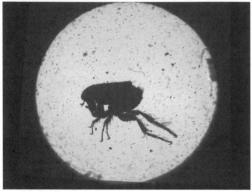

FIGURE 2.8
Dollond solar microscope slides. Projection of fleas with Dollond's instrument. Flea from Junker's sample of preparation (top left), flea prepared by the technician from the LMU (top right), and flea prepared by the author (bottom left). Courtesy Peter Heering.

order to dehydrate and thus preserve the specimen. Moreover, the flea was also embedded in entellan. To the best of my knowledge, both techniques were not used in the eighteenth century. Thus the image of this specimen can be taken as an indication of the potential of the instrument but does not show what people in the eighteenth century may have seen.

Dynamic Projections

These projections are not the only ones that were carried out with solar microscopes; another area that was extremely popular could be labeled with the term "dynamic projections." These projections were either made with water from hay infusions or with salty water. The former were used to demonstrate very minute living organisms that were to be found in the liquid. Equally popular were projections with salty water, as the heat of the sunlight accelerated the evaporation and thus promoted the growth of crystals. This was considered to be one of the most appropriate uses of solar microscopes and showed the major advantage of the instrument: "That the configurations [of salts] due to the heat of the focal point occur very apace and to the greatest delight of the spectators. Which is without any backtalk one of the pleasantest observations that can be made with a solar microscope."[26]

Like with the attempt to make my own preparations it is not clear how these projections could be replicated. Some instruments contained particular sliders that were supposed to be used in projections of liquids. However, none of these sliders has survived for the instruments held at the Deutsches Museum. Thus, it was decided to place small drops of potassium nitrate solution on a glass disc fixed in the aperture of a slider. This slider was immediately placed in the solar microscope. On the screen the drops could be seen easily (figure 2.9), some of them even acting like a lens. After some five minutes the beginning of the crystallization could be observed.[27] At first, smaller crystals are seen to be formed in the drop; these crystals can float in the liquid. As less and less liquid remains, the structure of the crystals becomes larger until the water has completely vanished and only the crystals remain. Although this is a dynamic process, the formation of the crystals is not continuous, particularly at the beginning of this process only a very quick growth can be observed. In spite of this discontinuous aspect of crystal growth the pictures are again misleading: Once this process has started, the appearance of the event is dynamic, thus the situation is more dynamic than the pictures and descriptions might indicate. Apart from that, some colors appear on the screen, mostly due to aberration effects in the liquid. The crystals themselves were colorless and appeared as shadowy objects on the screen, although some of them were semitransparent during their formation. However, some copperplates published by Ledermüller show colored crystals, the formation of which he claimed to have observed with the solar microscope.[28]

Conclusion

What can be learned from this study on solar microscopes? First of all it has become clear that the quality of the instruments can differ significantly with respect to the quality of the optical components as well as convenience of handling. But even Junker's cheap instrument—which can be interpreted as not being technologically that sophisticated—can be used for producing remarkable projections. However, some inconveniences in working with this device correspond to texts from the eighteenth century: Complaints that the image is always moving when the mirror is readjusted become understandable after working with Junker's solar microscope. On the other hand, some changes in the technical realization of the principle of the solar microscope have become understandable. Dollond's instrument seems to be stabilized in a manner that even unskilled persons are able to produce successfully impressive projections within a short amount of time. In this respect, a lot of the skills necessary to use Junker's solar microscope appropriately can be regarded as being materialized in the Dollond instrument. Although this device seems to be superior, there is also a drawback: When working with Dollond's microscope and wooden sliders—either from Junker's set or newly prepared ones— it turned out that these sliders neither fitted well into the instrument nor were they illuminated completely. This made it necessary to move the slider up and down in the instrument, a procedure that was far more difficult than in case of Junker's solar microscope. Consequently, it can be said that the materialization of skills in the case of Dollond's instrument led also to a lack of flexibility once other sliders than those accompanying the instrument were used.

Another aspect that became clearer during the study is related to several accounts of the eighteenth century in which the use of solar microscopes was praised: "The sun's rays being directed by the looking glass through the tube upon the object, the image or picture of the

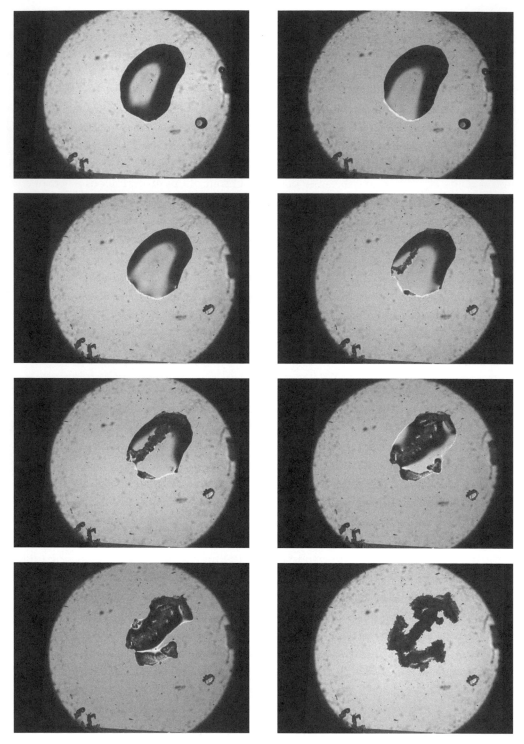

FIGURE 2.9
Solar microscope slides, crystallization of potassium nitrate, crystallization process starts with the top-left image, then top-right, and so on, in sequence to full crystallization. Courtesy Peter Heering.

object is thrown distinctly and beautifully upon a screen of white paper, and may be magnified beyond the imagination of those who have not seen it."[29]

Having seen and demonstrated projections, it became understandable why Priestley and others used such descriptions. The aesthetical attractiveness cannot be described although it is very visible in the darkened room. However, in this respect I am very much in the situation of an eighteenth-century natural philosopher: I cannot communicate my results (with respect to the aesthetics of the projections) through written publications. Only by inviting people into the laboratory is it possible to make these experiences available to other people. In this respect it is not only the size of the images—although the one of a flea with a length of some four meters is certainly impressive. It is the entire situation that contributes to the aesthetical impression, being in a darkened room and looking at clear, bright images is very different to any microscopic experience made with other devices.

What has become obvious in respect to the quality of projections is the importance of the preparations. However, as it was even possible for me to make preparations that appeared to be satisfactory, it can be questioned whether this had been a major problem in the eighteenth century. On the other hand, it is not clear what the criteria for a good preparation (or a good image on the screen) might have been. Thus it could be possible that what I nowadays consider as being a good preparation might have been unsatisfactory in the eighteenth or nineteenth centuries. Changing criteria for what is considered an adequate microscopic image could thus have resulted in the nineteenth century as criticisms of solar microscopes. However, these criticisms were not limited to this aspect; it was also pointed out very clearly that projections with solar microscopes were considered unscientific. As already discussed, this stood in contrast to the eighteenth-century notion of the solar microscope. One aspect that might be helpful in understanding this change becomes obvious when the instrument is used in a salon-like situation, that is, the projections are shown to a group of persons. Here it gets clear that through collective viewing of the projections, spectators started to discuss what they were seeing. This is a situation that is very much like the accounts of eighteenth-century salon discourses that were part of experimental practice.

However, toward the end of the eighteenth century a different style of experimentation became established that stood in contrast to these enlightenment demonstrations: scientific practice became professionalized, one aspect being that experiments were carried out in specific rooms where no visitors were admitted and the results were communicated in a mathematical form. These criteria of this newly developed style of experimentation were not met by experiences made with the solar microscope, consequently these could not be considered as being scientific experiments. Thus the criticism of authors such as Goring or Marbach can be seen as an indication of a new understanding of scientific practice.[30]

Notes

This manuscript was prepared in 2004 while I was in the Scholar in Residence Programme of the Deutsches Museum, Munich. For a detailed interpretation of the practice with the solar microscope see Heering, "The Enlightened Microscope"; for an account of the experiences, see Cavicchi, "A witness account." I would like to thank the research institute of the Deutsches Museum for offering me excellent working conditions. In particular, I would also like to thank Ulf Hashagen and Christian Sichau for discussions during my stay in Munich.

1. Gehler, *Physikalisches Wörterbuch*, 100.

2. Ibid.

3. Goring is quoted in Bradbury, *The Evolution of the Microscope*, 159. It should be remarked that Goring's statement is an example for strategies described by Catherine Wilson: "The feminization of the microscope assisted in reducing the prestige of an instrument of interest to and usable by ladies." Wilson, *The Invisible World*, 228.

4. Brisson, *Traité élémentaire ou Principes de Physique*, 510.

5. Marbach and Cornelius, *Physikalisches Lexikon*, 1070.

6. I am indebted to Klaus Staubermann who made the solar microscopes in the Utrecht collection accessible to me; Jan Deiman supplied me with the optical data of these instruments. The information regarding the solar microscopes kept in London and Oxford can be found in the Online Register of Scientific Instruments.

7. Dollond offered a solar microscope on a price list that had been printed between 1780 and 1793 for the price of £5 15s. 6d.; a similar price is given on a price list that was produced between 1805 and 1820. Other instrument makers in London were slightly cheaper with their instruments: Adams offered solar microscopes at £5 5s. 0d. in 1784. This price is also given in a list of instruments offered by Nairne and Blunt that was published together with a leaflet signed by Nairne.

8. According to the description Junker published, there are still some parts missing: a second microscope lens and some sliders that were meant to show living organisms in a drop of water. As the tube of the first solar microscope did not fit into the holder of the second one, a new tube had to be made. This tube was—like the original one—made from cardboard.

9. Junker mentions his motivation for producing solar microscopes in a short leaflet that was published in January 1791 and republished in September 1791. According to this leaflet, an English solar microscope cost 50–100 Thaler. Junker offered his instrument in the January leaflet for 5 Thaler and in the September edition for 7 Thaler (see Junker, *Ueber das Sonnen-Microscop*).

10. Ledermüller, *Nachlese seiner Mikroskopischen Gemüths- und Augen-Ergötzung*, 44. Ledermüller's difficulty resulted from the fact that his solar microscope suffered from the problem that by turning the mirror the slider also moved. Consequently he had to finish his sketch before readjusting the mirror.

11. Baker, *The Microscope Made Easy*, 23f.

12. It has to be remarked that the range of the depth of sharpness is very small, thus, as soon as objects are three-dimensional, it is only possible to have parts of it well-focused.

13. The sliders corresponded to what Bracegirdle describes as eighteenth-century standard: "Most preparations made until the end of the eighteenth century were sliders, an ivory or bone strip with several openings, each containing two discs of mica held in place with a brass circlip, and the object mounted between them." Bracegirdle, "Looking Again at Old Microscope Slides," 109. The only unusual aspect in this respect is Junker's sliders being made from wood; however, this is understandable from his manufacturing background.

14. Hacking, *Representing and Intervening*, 192. Hacking's notion of the microscope being a toy for ladies and gentlemen seems to miss a point: This image is mainly developed in the nineteenth century; yet, as I will argue, experimental practice and related criteria for what had been considered as being scientific changed between the eighteenth and nineteenth centuries. Thus, criticisms of the

nineteenth century were related to the standards of that period and cannot be seen as an adequate description of the eighteenth-century situation.

15. Nairne, *Description and Use of the Compound Microscope.*

16. This kind of preparation seems to be fairly typical for eighteenth-century microscopes: "Favourite specimens for mounting were shavings of timber, which look pretty and impressive when magnified about 40 times; parts of insects (or whole insects, there being no shortage of fleas and the like); lace and other textiles; and similar objects." Bracegirdle, "Looking Again at Old Microscope Slides," 109.

17. This does not correspond to Bracegirdle's account: "Quite a variety of objects is usually present on one slider, the only common factor being that they all needed the same magnification as the slider was slid across the stage under the objective." Bracegirdle, "Looking Again at Old Microscope Slides."

18. This is a result of the instrument and not of the sliders as I have also made projections using the Dollond instrument with Junker's sliders and vice versa.

19. Before experimenting with solar microscopes, all sliders were examined with an ordinary light microscope and the specimens appeared already colorless.

20. In the context of reconstructing an apparatus for analyzing an experiment with the replication method, Christian Sichau has pointed out that apparatus change over time and thus an instrument kept in a museum cannot be taken as being in the same condition it had been during its original use. For a detailed discussion of the possible temporal instability of setups see Sichau, *Die Viskositätsexperimente von J. C. Maxwell und O. E. Meyer.* See also Heering, "Weighing Heat."

21. Wiedeburg, *Beschreibung eines verbesserten Sonnen-Microscops,* 10.

22. Baker, *The Microscope Made Easy,* 56.

23. I am indebted to Heidrun Schöl (Institute of Comparative Tropical Medicine and Parasitology, Ludwig-Maximilian University, Munich) who prepared a flea, a louse, and some mites for me.

24. Daumas, *Scientific Instruments of the 17th & 18th Centuries and Their Makers,* 236.

25. Anonymous. "Nachricht von einem brauchbaren und wohlfeilen Sonnen-Mikroskop," 87.

26. von Gleichen, *Auserlesene mikroskopische Entdeckungen bey den Pflanzen, Blumen und Blüthen,* 154.

27. It has to be remarked that these experiments were carried out in October; thus it might be possible that the time necessary for the beginning of the crystals' formation was less during the summer.

28. I did not use only potassium nitrate, but also ammonium nitrate, ammonium chloride, and copper sulfate. Although the last appears in a colorful blue, the crystals appear dark in the projections. According to Ledermüller he used—among others—a solution of verdigris. It is not clear whether the colors might be a result of impurities of the chemicals he used. Besides, the relevance of this kind of projection can also be inferred from figure 2, which Ledermüller published to illustrate the use of the solar microscope. The image projected onto the wall results also from the formation of salt crystals.

29. Priestley, *The History and Present State of Discoveries Relating to Vision, Light, and Colours,* 742.

30. I am using the concept of style of experimentation with reference to Ludwik Fleck's epistemological concept. On the style of experimentation developed toward the end of the eighteenth century see Heering, *Das Grundgesetz der Elektrostatik. Experimentelle Replikation und wissenschaftshistorische Analyse.*

References

Anonymous. "Nachricht von einem brauchbaren und wohlfeilen Sonnen-Mikroskop." *Magazin für das Neueste aus der Physik und Naturgeschichte* 7, no. 3 (1791): 84–87.

Baker, Henry. *The Microscope Made Easy.* 5th ed. London, 1769. Reprinted, Lincolnwood: Science Heritage Limited, 1987.

Bracegirdle, Brian. "Looking Again at Old Microscope Slides." *Endeavour* N. S. 18 (1994): 109–14.

Bradbury, Savile. *The evolution of the microscope.* Oxford; New York, Pergamon Press, 1967.

Brisson, Mathurin-Jacques. *Traité élémentaire ou Principes de Physique*, vol. 2. Paris, 1789.

Cavicchi, Elizabeth. "A witness account of solar microscope projections: collective acts integrating across personal and historical memory." *British Journal for the History of Science* 41 (2008): 369–83.

Daumas, Maurice. *Scientific Instruments of the 17th & 18th Centuries and Their Makers.* Translated and edited by Mary Holbrook. London: Portman Books, 1989 (1st English ed. New York: Praeger Publishers, 1972).

Gehler, Johann Samuel Traugott. *Physikalisches Wörterbuch.* Vierter Teil. Leipzig, 1791.

Hacking, Ian. *Representing and Intervening: Introductory Topics in the Philosophy of Natural Science.* Cambridge: Cambridge University Press, 1983.

Heering, Peter. *Das Grundgesetz der Elektrostatik. Experimentelle Replikation und wissenschaftshistorische Analyse.* Wiesbaden: Deutscher Universitäts-Verlag, 1998.

Heering, Peter. "The Enlightened Microscope—Re-enacting and Analysing Projections with 18th Century Solar Microscopes." *British Journal for the History of Science* 41 (2008): 345-67.

Heering, Peter. "Weighing Heat: The Replication of the Experiments with Lavoisier's and Laplace's Ice-Calorimeter." In *Lavoisier in Perspective*, edited by Marco Beretta. Munich: Deutsches Museum, 2005, 27–41.

Junker, F. A. *Ueber das Sonnen-Microscop.* Magdeburg, January 1791, 2nd ed.

Ledermüller, Martin Frobenius. *Nachlese seiner Mikroskopischen Gemüths- und Augen-Ergötzung.* Nürnberg, 1762.

Marbach, Oswald, and C. S. Cornelius. *Physikalisches Lexikon.* 2nd ed., vol. 4. Leipzig: Otto Wiegand, 1856.

Nairne, Edward. *Description and Use of the Compound Microscope, as Made and Sold by Edward Nairne, At Number 20, in Cornhill, opposite the Royal Exchange, London.* n.p. and n.d.

Online Register of Scientific Instruments. www.isin.org/ressrch.asp?All=solar_microscope (accessed October 27, 2004).

Priestley, Joseph. *The History and Present State of Discoveries Relating to Vision, Light, and Colours.* London: J. Johnson, 1772.

Sichau, Christian. *Die Viskositätsexperimente von J. C. Maxwell und O. E. Meyer: Eine wissenschaftshistorische Studie über die Entstehung, Messung und Verwendung einer physikalischen Größe.* Berlin: Logos, 2002.

von Gleichen, Wilhelm Friedrich Freiherr. (genannt Rußworm). *Auserlesene mikroskopische Entdeckungen bey den Pflanzen, Blumen und Blüthen, Insekten und anderen Merkwürdigkeiten.* Nürnberg, 1777.

Wiedeburg, Joh. Ernst Basilius. *Beschreibung eines verbesserten Sonnen-Microscops.* Nürnberg, 1758.

Wilson, Catherine. *The Invisible World: Early Modern Philosophy and the Invention of the Microscope.* Princeton, NJ: Princeton University Press, 1995.

Instruments
in Industry

CHAPTER 3

Refractometers and Industrial Analysis

Deborah Jean Warner

ODERN INDUSTRY is hooked on instruments, using them to monitor manufacturing processes and assess the quality of finished products. Historians, however, have largely overlooked these practical devices in favor of more attractive instruments of brass and glass. This essay, a modest attempt to redress the balance, focuses on the refractometer, an instrument that measures the refractive index of substances. Like many instruments of industrial analysis, the refractometer was appreciated by managers of rapidly expanding businesses and by scientists working in industrial, governmental, and academic laboratories. If artisans with refined sensory perceptions protested this instrumental intrusion into their work, their voices have not been heard. Most of my evidence comes from the United States, but there is no reason to believe that the American experience differed, in any essential respect, from that of other industrialized nations.

As with so many technologies, it is difficult to identify the first refractometer. I begin this story with Ernst Abbe, a mathematical physicist at the University of Jena who was appointed research director in Carl Zeiss's Optische Werkstätte in 1866 and was asked to put the design and manufacture of optical instruments on a more scientific basis. Abbe developed his first refractometer in 1869, in conjunction with an effort to produce optical glass of uniform refractive index.[1] After realizing that the refractive index of the immersion medium was a critical factor in the definition of the numerical aperture of a lens system, Abbe used the refractometer to develop immersion objectives for microscopes. Zeiss began marketing refractometers in 1881, noting that they were "occasionally employed in microscopical investigations."[2] Abbe became Zeiss's partner in 1876 and head of the firm following Zeiss's death in 1888. Two years later, having recognized the growing potential of the refractometer market, Zeiss established a department of optical measuring instruments. Led by Carl Pulfrich, a "physicist of practical experience," this department developed several new forms.

The first industrial use of the refractometer outside the Zeiss factory pertained to oleomargarine, an inexpensive beef-fat spread that had been mistaken for butter ever since commercial production had begun in 1873. The American Chemical Society discussed oleomargarine in 1881, noting that the sale of this substance "has become so extensive in this country, that a purchaser of butter is never sure whether he is getting true butter or its imitation."[3]

American chemists knew that the problem was not unique to the United States. In 1886 *Science* reported that the Imperial Health Office in Berlin had found that butter and oleomargarine "cannot, in most cases, be distinguished from each other by their external appearance, or by the senses in any way, without the aid of physical or chemical investigation."[4] Harvey Wiley, a man of enormous energy and ego who had studied in Germany in the late 1870s and who remained in contact with his European colleagues even after becoming chief chemist in the U.S. Department of Agriculture (USDA), reported in 1887 that a German chemist had found the refractive index of butter fat to be lower than that of other glycerides.[5] By March 1888, Wiley was using an Abbe refractometer in a "series of experiments to determine the refraction of different oily substances." There being no literature on the subject, he believed this work would be "an interesting contribution to the present knowledge of the qualities of fats.[6] Moreover, since there were many "professional chemists" in the United States who devoted themselves to food analysis and other sorts of "private work," Wiley knew that he was addressing a large and growing audience. The Association of Official Agricultural Chemists, over which Wiley had enormous influence, recommended the refractometer as an official instrument of analysis in 1895.[7] The earliest refractometer in the Smithsonian collections—marked "N° 148 Carl Zeiss Jena"—was delivered to J. W. Queen & Company in Philadelphia in 1890 and sold to the University of Michigan.

Zeiss showed several refractometers at the Columbian Exposition held in Chicago in 1893 and discussed them at length in *Optical Measuring Instruments*, an illustrated text available in either English or German. Addressing the community of "pharmaceutical and manufacturing chemists and others engaged in compounding substances," Zeiss noted that refractometers could be used "to distinguish many substances and to ascertain their degree of purity (adulteration of victuals), or to determine the percentage or concentration of many solutions and mixtures." And, unlike some instruments, refractometers required no knowledge of optics or any delicate procedures for their manipulation.[8]

One of the new Zeiss refractometers, the butyro, was similar to the Abbe original but simpler, more robust, less expensive, and especially suited for fats and oils.[9] Eimer and Amend, an important instrument firm in New York, began advertising the butyro in 1895. Wiley discussed it in his influential *Principles and Practice of Agricultural Analysis* (1897). Leffmann and Beam mentioned it in their *Select Methods in Food Analysis* (1901), noting that it "has been strongly recommended for the examination of butter" and was "equally adapted for the examination of fats and oils." Albert Leach, a chemist with the Department of Food and Drug Inspection of the Massachusetts State Board of Health, featured the butyro in his *Food Inspection and Analysis* (1904), a book written "for the use of public analysts, health officers, sanitary chemists, and food economists." It was, he said, "by far the most useful and convenient form of the instrument for the food laboratory."[10]

As chemists gained familiarity with refractometers, they began developing new proce-dures for their use. Two chemists employed by the USDA developed a method of correct-ing the butyro readings for temperature.[11] Another prepared two tables—one showing the "Conversion of Butyro-Refractometer Readings to Indices of Refraction" and another show-ing the "Temperature Correction for Refractive Indices of Oils"—for the 1907 edition of *Van Nostrand's Chemical Annual*.[12] Working with his colleague, Hermann Lythgoe, Leach devised a slide rule for converting the arbitrary numbers found on butyro refractometers into actual indices of refraction.[13]

Americans also used butyro refractometers in new but related areas. Carlos Cochran, a chemist with the Dairy and Food Commission of the Pennsylvania Department of Agricul-ture, was using a butyro to detect foreign fats in lard and butter. W. H. Hess and R. E. Doolittle, of the Michigan Dairy and Food Department, used a butyro to detect "processed" butter (that is, rancid butter that had been renovated). Lythgoe used a butyro to analyze cas-tor and cod-liver oil. By 1921, American dealers were claiming that the butyro refractome-ter was being used with "cheese, margarine, cocoa butter, lard and other comestible fats, salad oil, cod-liver oil, lubricating oils, alkalies, linseed oil, varnish, turpentine, petroleum, paraf-fin, ceresin and other forms of wax."[14]

The butyro, it must be said, was not the first specialized refractometer on the market. That honor goes to the oleorefractometer, an instrument designed by Émile Hélaire Ama-gat, a physicist, and Ferdinand Jean, the director of the laboratory at the Bourse de Com-merce, described to the Académie des Sciences in 1889, and produced by P. Pellin, a leading instrument maker in Paris. The oleorefractometer was a differential apparatus that allowed the rapid determination of the refractive difference of two oils under the same conditions.[15] The refractometer designed by Charles Féry was also suitable for analyzing oils and fats—Pellin referred to it as an industrial instrument—as was the liquiscope devised by M. Son-den of Stockholm around 1891.[16]

The immersion (or dipping) refractometer, designed by Pulfrich and introduced in 1899, was an Abbe refractometer of short range in which the prism dips into the sample to be ana-lyzed. Visitors to the Zeiss display at the World's Fair held in St. Louis in 1904 were told that it was useful "for rapidly determining the concentration of solutions" such as "the extract and the alcohol in beer."[17] Leach and Lythgoe used an immersion refractometer to deter-mine the refractive indices of ethyl and methyl alcohols. So important was this work that Leach's obituary called attention to his "refractometric method of detecting wood alcohol in liquor [which] was devised at the time when this adulterant was causing havoc among the consumers of cheap liquors and at once superseded the laborious and unsatisfactory meth-ods then in vogue."[18]

Robert Schwarz, a chemist affiliated with the First Scientific Station for the Art of Brew-ing in New York, argued in 1913 that European brewing chemists were generally agreed "that the results obtained with the refractometer are perfectly reliable for factory control" and that Americans should follow their lead.[19]

Leach, who argued that the immersion refractometer had "many features" that "espe-cially commend it to the use of the food analyst," used one to detect milk that had been fraudulently watered, a common problem in this period.[20] Willard Bigelow, director of research at the National Canners Association, and his colleague, Fred Fitzgerald, used one to determine the refractive index and specific gravity of fresh and canned tomatoes and tomato pulp. By 1913, the authoritative *Dictionary of Applied Chemistry* could claim that the immersion refractometer had led to a "considerable extension of refractometric methods in chemical work."[21]

The Abbe refractometer with heatable prisms was especially suited for the analysis of sugar, a substance that has long been a major element of the American diet and economy. Two USDA chemists determined the percentage of dry substance in various sugar solutions with an instrument of this sort and found that the refractometric analysis of sugar gave the same results as did the specific gravity method but had the advantage of "speed, ease of manipulation, and amount of sample required for the determination." Another USDA chemist thought that this instrument should appeal greatly to "the technical chemist of a sugar house" on account of its ease of manipulation and accuracy compared with other methods.[22] Samuel S. Peck, a chemist with the Experiment Station of the Hawaiian Sugar Planters Association, found that the difficulty of determining the total solids of molasses "has been largely if not completely solved by the discovery of the applicability of the refractometer for this purpose."[23] Charles Albert Browne, a chemist with impressive scientific credentials, included lengthy discussion of the Abbe refractometer in his *Handbook of Sugar Analysis* (1912), "a practical and descriptive treatise for use in research, technical and control laboratories."[24] H. C. Prinsen Geerlings echoed these conclusions in his *Chemical Control in Cane Sugar Factories* (1917), a book originally published by the Hawaiian Chemists' Association.[25]

Citing the wide adoption of the refractometric method in sugar factories, Zeiss introduced its sugar refractometer in 1911. This was a robust Abbe instrument with a relatively narrow refractive index range and a scale for reading the direct percentage of sugar in a solution. Some Féry refractometers, at this time, were calibrated for reading directly the percentage of sugar in aqueous solutions.[26] A German chemist argued in 1921 that sugar refractometers "should certainly be found in every sugar laboratory." These instruments were "simple and accurate" and able to "read easily up to 95% total solids." The "figures for dry substances obtained by its use" might not agree "with the total solids by drying nor with those by spindling," he admitted, but "as long as only comparative results are needed, like in the sugar house, it makes no difference which method is used, as long as it is the only one employed." A Russian chemist called attention to the bottom line: if each sugar factory would obtain a refractometer and establish "an ideal boiling diagram," they would find that "even a non-skilled worker may be entrusted with the job of boiling."[27]

The Pulfrich refractometer, introduced in the mid-1890s, was "chiefly used for investigations of a purely scientific nature" but practical applications were not unknown.[28] Two American chemists analyzed petroleum and solid hydrocarbons with an instrument of this sort as early as 1902.[29] By 1913, Zeiss was making a gas interferometer "for ascertaining the difference between the refractive indices of a given gas and a standard gas."[30]

Arthur Lyman Dean and Ernest Bateman, both with the U.S. Forest Service, used a refractometer to distinguish coal tar and coke oven creosotes from blast furnace and water gas creosotes—a matter of some import in the development of preservatives suitable for such wooden items as telephone poles and railroad ties.[31] Ernest J. Parry, an analytical and consulting chemist in London, mentioned the refractometer in connection with essential oils in 1899, and in 1908 he described the Abbe design as being "the most useful" for "for ordinary work."[32] The Larkin Company in Buffalo, New York, one of the largest soap manufacturers in the United States, reported that the refractive index of liquid soaps was directly proportional to their total solids content.[33] Chemists working under the auspices of Schimmel and Company, a large perfume manufacturer in Militz, near Leipzig, used the refractometer on volatile oils.[34] In physiological and biochemical laboratories, refractometers were used to determine the quantity of sugar in urine and albumen in blood.[35]

One of the few German instrument makers who challenged Zeiss's dominance of the refractometer market was Hans Heele, in Berlin, whose Abbe-type refractometer was discussed at the Society of Chemical Industry meeting in May 1909.[36] The Great War severed access to German firms and forced Britains and Americans to begin making those instruments that had become necessary for military and industrial purposes. Adam Hilger, Limited introduced its Zeiss-type instruments in 1918, noting that they incorporated several important improvements,[37] and that they were "continually finding new applications" in such industries as butter, edible and technical fats, oils, waxes, sugars, syrups, essential oils, glue, gelatine, petroleum, paint, varnish, gas, tanning, brewing and distilling, margarine and drug extraction.[38] American advertisements noted that Hilger's immersion refractometer "finds frequent application" in food and drug laboratories, breweries, distilleries, and medical laboratories.[39] The Bausch & Lomb Optical Company in Rochester, New York, and the Spencer Lens Company in Buffalo, New York, began making similar instruments and publishing similar advertisements in 1922.[40] All of these firms, along with Zeiss, were still making optical refractometers in the 1950s.

Zeiss introduced refractometers for process control in the post-war period.[41] The interference refractometer designed by Fritz Haber, the German-Jewish chemist who won the Nobel Prize in 1918, was used to measure the concentration of dilute solutions, by U.S. Chemical Warfare Service at Edgewood Arsenal.[42] The Zeiss Works refractometer, available by 1926, served "for controlling the quality of a solution while boiling in a closed vessel or while flowing through a pipe or other channel, for example with reference to its water content." It did away with the "necessity of extracting a sample" and "stood its first trials in the service of the sugar industry."[43]

Tables of refractive indices provide yet another handle on the adoption of refractometry. Bernhard Wagner, a graduate student in chemistry at the University of Jena, used an early Zeiss immersion refractometer to measure the strength of various solutions and show that refractometric results compared favorably with those based on specific gravity. Zeiss published Wagner's tables in 1907 and distributed them to chemists around the world. Carl Zeiss, Jena, brought out a fourth edition in 1955.[44] Describing Wagner's tables as "the most important contribution to the technical use of the refractometer for aqueous solutions," Adam Hilger, Limited published a table of the refractive indices of essential oils, another of oils, fats, and waxes, and a third of sugar.[45] The first volume of the *International Critical*

Tables of Numerical Data, Physics, Chemistry and Technology, prepared by the U.S. National Research Council, included refractive index and dispersion among the important physical properties of chemical substances; the seventh volume includes a table of the refractivity of all gases and vapors and of elementary substances in the isotropic solid and liquid states.[46]

One might also look at the accounts of refractometry that appeared in general books on chemical technology and specialized texts such as D. Sidersky, *La réfractometrie et ses applications pratiques* (1909) and Walter Adolf Roth and Fritz Eisenlohr, *Refraktometrisches Hilfsbuch* (1911), as well as the apparatus used in the increasing number of classes of practical, industrial, or general chemistry. Finally, it would be interesting to tease out the various meanings that a high-tech instrument, such as a refractometer, had for the chemists who used them in their daily work and for the consumers who bought products tested with them. My rough examination shows that the issues raised in the early decades of practical refractometry would continue throughout the twentieth century.

Conclusion

This issue of *Artefacts* raises several questions about scientific instruments, all of which pertain to refractometers. One, the question of definitions, can best be answered historically and culturally. While academic scientists generally use the term "scientific instrument" in connection with the devices they use in their laboratories, historians of instruments have focused much of their attention on optical and mathematical devices that were designed for practical rather than philosophical purposes. When *Instruments of Science. An Historical Encyclopedia* was in the planning stage, Robert Bud insisted we include instruments that were designed for quality and process control in the real world, both those that emerged from science and those that did not. This exploration of refractometers obviously follows that agenda.

Another question pertains to the use that historians might make of instruments, especially those housed in museum collections. Here I readily admit to some skepticism. Object analysis might be appropriate in those instances in which the instrument itself is the primary, if not the only, available text. But it seems pointless with regard to modern instruments for which we have extensive and intelligent literature prepared by manufacturers, dealers, and users. And, since many of these modern instruments are fairly delicate, and many were out of alignment long before they came into museum collections, any effort to make observations with them would be fraught with difficulties, if not downright silly. In the absence of compelling evidence to the contrary, I am willing to believe the German-English chemist Julius Lewkowitsch, who in 1904 called attention to the "ease and rapidity with which the refractive index can be determined, in consequence of the recent improvements in optical apparatus."[47]

For most of us who are not practicing chemists, refractometers are black boxes that, even when opened, are not especially intelligible. Not knowing why we should look at these devices, or what features of them we should focus on, we concentrate on the persons by whom, and the contexts in which, these instruments were made and used. And that gets us to the third question of this volume—what were the cultural and other contexts of instruments? To this end, I suggest we approach instruments by way of life histories and performance characteristics. The former path leads to questions about instruments' invention, production, distribution, and use. The latter leads to questions about instruments' cost (of production, maintenance, ancillary resources, etc.), skills needed (for production and use), reliability, symbolic meanings, and so forth.

Finally, I would refer to Robert Multhauf, my first boss at the National Museum of American History (then the Museum of History and Technology), who taught us that museum objects were, if nothing else, valuable as stimuli for research. Anyone might tackle the history of the refractometers, but it's the curators who collect, catalog, and dust these things who are most likely to do so.

Notes

1. Paselk, "The Evolution of the Abbé Refractometer." Abbé, *Neue Apparate.*

2. Zeiss, *Microscopes and Microscopical Accessories*, 47–48; similar text apparently appeared in the 1881 edition of this work.

3. Casamajor, "Detection of Oleomargarine."

4. Anonymous, "Comment and Criticism."

5. Wiley, "Dairy Products"; Skalweir, "On Butter Testing"; Moore, "Distinguishing Between Natural and Artificial Butter."

6. Anonymous, "Washington Scientific News." Wiley, "Lard and Lard Adulterations."

7. Wiley, "Report of the Chemist"; Wiley, "Methods of Analysis."

8. Zeiss, *Optical Measuring Instruments*, preface and pp. 8–9. Pulfrich, "Ein neues Refractometer"; Pulfrich, *Das Totalreflectometer.*

9. Zeiss, *Optical Measuring Instruments*, 2–14; Zeiss, *Das Butterrefraktometer*; Zeiss, *Tabelle zum Zeiss Butterrefraktometer.*

10. Eimer & Amend, *Illustrated Wholesale Catalogue*, 339–42; Leffmann and Beam, *Select Methods*, 158; Leach, *Food Inspection*, 389; the third edition (1913) devoted a whole chapter to the refractometer.

11. Tolman and Munson, "Refractive Indices."

12. Olsen, *Van Nostrand's Chemical Annual*, 64–65.

13. Leach and Lythgoe, "A Comparative Refractometer Scale," and "Improved Refractometer Slide."

14. Cochran, "The Detection of Foreign Fats"; Hess and Doolittle, "Methods for the Detection"; Lythgoe, "The Optical Properties"; Thomas, *Laboratory Apparatus*, 520.

15. Amagat and Jean, "Sur l'analyse optique"; Jean, "Applications of the Oleorefractometer"; Muter, "Methods and Apparatus."

16. Féry, "Sur un nouveau réfractom?etre," and "Réfractom?mtre ? cuve chauffable." Pellin, *Instruments d'Optique*, 10–11. Thorpe, *A Dictionary of Applied Chemistry*, 554–55. The liquiscope is noted in *Science* 18 (1891): 201.

17. Pulfrich, "Über das neue Eintauchrefraktometer," 206.

18. Leach and Lythgoe, "The Detection and Determination"; and *Science* 22 (1905): 78; anonymous, "Albert Ernest Leach."

19. Schwarz, "The Use of the Immersion Refractometer." Schwarz mentions Tornoe's method using sodium light and a specially constructed type of differential refractometer, noting that this method, while fairly rapid and quite accurate, did not find wide application. Eimer & Amend, *Illustrated Catalogue*, 325, lists a Tornoe refractometer manufactured by Franz Schmidt & Haensch.

20. Leach and Lythgoe, "The Detection of Watered Milk"; Lythgoe and Nurenberg, "A Comparison of Methods."

21. Bigelow and Fitzgerald, "The Relation of the Refraction," and "Refractometer," 553.

22. Tolman and Smith, "Estimation of Sugars"; Bryan, "The Estimation of Dry Substance."

23. Peck, "Total Solids in Mill Products."

24. Browne, *A Handbook of Sugary Analysis*, chapter 4. "C. A. Browne" in *Dictionary of American Biography*.

25. Geerlings, *Chemical Control*, 22–23; this is the 1916 revised and enlarged edition of *Methods of Chemical Control for Cane Sugar Factories*.

26. Zeiss, *Ein neues Refraktometer*; Eimer & Amend, *Chemical and Metallurgical Laboratory Supplies*, 354; Scientific Materials Co., *General Apparatus Catalogue*, 422.

27. Herzfeld, "Can the Refractometer be Recommended?"; Kartashev and Savinov, "The Application of the Industrial Zeiss Refractometer."

28. Pulfrich, "Ein neues Refraktometer." Thorpe, *A Dictionary of Applied Chemistry*, 550–51.

29. Mabery and Sheperd, "A Method for Determining the Index of Refraction."

30. Eimer & Amend, *Chemical and Assay Laboratory Supplies*, 357.

31. Dean and Bateman, "The Analysis and Grading of Creosotes."

32. Parry, *The Chemistry of Essential Oils*, 83; and edition 102–7.

33. Hoyt and Verwiebe, "Determination of the Concentration of Liquid Soaps."

34. Gildemeister and Hoffmann, *The Volatile Oils*, 561–62.

35. Zeiss, *Refraktometer Abbéscher Konstrucktion*, and *Abbé's Refractometers*.

36. Lewkowitsch, "A New Refractometer"; Hans Heele, *New Refractometer*.

37. "The Abbé and Pulfrish [*sic*] Refractometers" concerned Hilger instruments shown at the British Scientific Products Exhibition. Simeon, "The Accuracy Attainable"; Twyman and Simeon, "Accuracy Control"; Stanley, "Improved Types of British Refractometers"; Hilger, *General*.

38. Hilger, *Instructions for Use*.

39. Central Scientific Co., *Catalog C-222*, 428–29.

40. Bausch & Lomb ad in *Journal of the Optical Society of America*; Bausch & Lomb *Chemical Engineering Catalog*, 324-325; and Bausch & Lomb, *Microscopes and Accessories*, 204-217. Spencer Lens Co., *Catalog of Spencer Products*; Spencer instruments were also available through Central Scientific Co., *Catalog C-222*, 427–31.

41. Löwe, "Das Löwe," *Optische Messungen*. Löwe was on the Zeiss payroll.

42. Macy, "Application of the Interference."

43. Zeiss, *Industrial Refractometer*, 1–2. Löwe, "Das Betriebsrefraktometer," xx–xx.

44. Wagner, *Tabellen zum Eintauchrefraktometer*. For distribution, see Zeiss, *Dipping Refractometer*, 3. Hilger is mentioned in Thomas, *Laboratory Apparatus*, 518.

45. Hilger, *Optical Methods*, 17. Goldsmith, *Tables of Refractive Indices*.

46. Washburn, *International Critical Tables*, esp. vol. 1, 165–74 and vol. 7, 1–19.

47. Lewkowitsch, *Chemical Technology*, 184.

References

Abbe, Ernst. *Neue Apparate zur Bestimmung des Brechungs- und erstreuungsvermögens fester und fluüsinger Körper.* Jena, 1874.

Amagat, E. H., and F. Jean. "Sur l'analyse optique des huiles et du beurre." *Comptes Rendus de l'Académie des Sciences* 109 (1889): 616–17.

Anonymous. "Albert Ernest Leach." In *National Cyclopaedia of American Biography*, vol. 19, 449–50.

Anonymous. "C. A. Browne." In *Dictionary of American Biography*, sup. 4, 113–15.

Anonymous. "Comment and Criticism." *Science* 7 (1886): 537–38.

Anonymous. "Proceedings of the Eighth Annual Convention of the Association of Official Agricultural Chemists." U.S. Department of Agriculture, Division of Chemistry, *Bulletin* 31 (1891): 201–2.

Anonymous. "Refractometer." In E. Thorpe, ed., *A Dictionary of Applied Chemistry*, vol. 4. London, 1913, 550–55.

Anonymous. "The Abbé and Pulfrish (sic) Refractometers." *Engineering* 106 (1918): 282–83.

Anonymous. "Washington Scientific News." *Science* 11 (1888): 102.

Bausch & Lomb. Advertisement. *Journal of the Optical Society of America* 6 (March 1922): inside front cover.

Bausch & Lomb. *Chemical Engineering Catalog* 7 (1922), Rochester, NY.

Bausch & Lomb. *Microscopes and Accessories.* Rochester, 1929.

Bigelow, W. D., and F. F. Fitzgerald. "The Relation of the Refraction, Specific Gravity and Solids in Tomatoes and Tomato Pulp." *Science* 41 (1915): 37–38.

Browne, C(harles) A(lbert). *A Handbook of Sugary Analysis.* New York, 1912.

Bryan, A. H. "The Estimation of Dry Substance by the Refractometer in Liquid Saccharine Food Products." *Journal of the American Chemical Society* 30 (1908): 1443–51; and *Science* 28 (1908): 185.

Casamajor, Paul. "Detection of Oleomargarine." *Journal of the American Chemical Society* 3 (1881): 83–87.

Central Scientific Co. *Catalog C-222.* Chicago, ca. 1922.

Cochran, C. B. "The Detection of Foreign Fats in Lard and Butter." *Journal of the American Chemical Society* 19 (1897): 796–99.

Dean, Arthur L., and Ernest Bateman. "The Analysis and Grading of Creosotes." *Circular of the U.S. Forest Service* 112 (1908): 102–7.

Eimer and Amend. *Chemical and Metallurgical Laboratory Supplies and Assayers' Materials.* New York, 1913.

Eimer and Amend. *Chemical and Assay Laboratory Supplies.* New York, 1913.

Eimer and Amend. *Illustrated Catalogue.* New York, 1910.

Eimer and Amend. *Illustrated Wholesale Catalogue of Chemical and Physical Apparatus.* New York, 1895.

Féry, Charles. "Réfractométre á cuve chauffable. Application á la mesure des corps gras." *Comptes Rendus de l'Académie des Sciences* 119 (1894): 332–34.

Féry, Charles. "Sur un nouveau réfractométre." *Comptes Rendus de l'Académie des Sciences* 113 (1891): 1028–30.

Geerlings, H. C. Prinsen. *Chemical Control in Cane Sugar Factories.* London, 1917.

Geerlings, H. C. Prinsen. *Methods of Chemical Control for Cane Sugar Factories.* Honolulu, 1916.

German Educational Exhibition. "World's Fair, St. Louis." *Scientific Instruments.* Berlin, 1904.

Gildemeister, E., and Fr. Hoffmann. *The Volatile Oils.* New York, 1913.

Goldsmith, J. N., ed. *Tables of Refractive Indices.* London, 1918–1921.

Heele, Hans. *New Refractometer for Investigating Solid and Liquid Objects.* Berlin, ca. 1914.

Hess, W. H., and R. E. Doolittle. "Methods for the Detection of 'Process' or 'Renovated' Butter." *Journal of the American Chemical Society* 22 (1900): 150–52.

Herzfeld, A. "Can the Refractometer be Recommended for Use in the Laboratory of the Sugar Factory?" *Chemical Abstracts* 15 (1921): 33913.

Hilger, Adam, Ltd. *Optical Methods in Control and Research Laboratories.* London, ca. 1920.

Hilger, Adam, Ltd. *Abbé Refractometer with Water Jacketed Prisms.* London, ca. 1918.

Hilger, Adam, Ltd. *General Catalogue of the Manufactures.* Section M. London, 1924.

Hilger, Adam, Ltd. *Instructions for Use of the Abbé Refractometer.* London, ca. 1918.

Hoyt, L. F., and Alma Verwiebe. "Determination of the Concentration of Liquid Soaps by the Immersion Refractometer." *Industrial and Engineering Chemistry* 18 (1926): 581–82.

Jean, F. "Applications of the Oleorefractometer of E. H. Amagat and F. Jean to the Detection of Sophistications." *Chemical News* 62 (1890): 296–97; abstract from *Bulletin de la Société Chimique de Paris.*

Kartashev, A. K., and B. G. Savinov. "The Application of the Industrial Zeiss Refractometer for the Study and Control of the Boiling of the Fill-Mass." *Chemical Abstracts* 21 (1927): 3280.

Leach A. E., and H. C. Lythgoe. "A Comparative Refractometer Scale for Use with Fats and Oils." *Journal of the American Chemical Society* 26 (1904): 1193–95.

Leach A. E., and H. C. Lythgoe. "Improved Refractometer Slide Rule and its Application." *Science* 24 (1906): 195.

Leach, A. E., and H. C. Lythgoe. "The Detection of Watered Milk." *Journal of the American Chemical Society* 24 (1904): 1195–203.

Leach, A. E., and H. C. Lythgoe. "The Detection and Determination of Ethyl and Methyl Alcohols in Mixtures by the Immersion Refractometer." *Journal of the American Chemical Society* 27 (1905): 964–72; and *Science* 22 (1905): 28.

Leach, Albert. *Food Inspection and Analysis.* New York, 1904.

Leffmann, H., and W. Beam. *Select Methods in Food Analysis.* Philadelphia, 1901.

Lewkowitsch, J. "A New Refractometer." *Journal of the Society of Chemical Industry* 28 (1909): 773–75.

Lewkowitsch, Julius. *Chemical Technology and Analysis of Oils, Fats, and Waxes.* 3rd ed. London, 1904.

Löwe, Fritz. "Das Betriebsrefraktometer." *Zeitschrift d. Ver. d. deutschen Zucker-Ind* 75 (May 1925).

Löwe, Fritz. "Das Refractometer im Fabrikslaboratorium." *Chemiker-Zeitung* 45 (1921): 25–27, 52–55.

Löwe, Fritz. *Optische Messungen des Chemikers und des Mediziners.* Dresden, 1925; 6th ed., 1964.

Lythgoe, H. C. "The Optical Properties of Castor Oil, Cod-Liver Oil, Neat's Foot Oil, and a Few Essential Oils." *Journal of the American Chemical Society* 27 (1905): 887–92.

Lythgoe, H. C., and L. I. Nurenberg. "A Comparison of Methods for the Preparation of Milk Serum." *Journal of Industrial and Engineering Chemistry* 1 (1909): 38–40.

Mabery and Sheperd. "A Method for Determining the Index of Refraction of Solid Hydrocarbons with the Pulfrich Refractometer." *Proceedings of the American Academy of Arts and Sciences* 38 (1902–3): 281.

Macy. Rudolph. "Application of the Interference Refractometer to the Measurement of the Concentration of Dilute Solutions." *Journal of the American Chemical Society* 49 (1927): 3070–76.

Moore, R. W., abstract of J. S. Kaliveit [sic]. "Distinguishing Between Natural and Artificial Butter by Means of the Refractometer." *Journal of the American Chemical Society* 9 (1887): 14.

Muter, Dr. "Methods and Apparatus in Use at the Laboratory of the Bourse de Commerce at Paris for the Analysis of Certain Commercial Organic Products." *Analyst* 15 (1890): 85–89.

Olsen, J. C., ed. *Van Nostrand's Chemical Annual.* New York, 1907.

Parry, E. J. *The Chemistry of Essential Oils and Artificial Perfumes.* London, 1899 & 1908.

Paselk, Robert A. "The Evolution of the Abbé Refractometer." *Bulletin of the Scientific Instrument Society* 62 (September 1999): 19–22.

Peck, S. S. "Total Solids in Mill Products by the Refractometer." *Report of the Experiment Station of the Hawaiian Sugar Planters Association* 27 (1908).

Pellin, Ph. *Instruments d'Optique et de Précision.* Fasc. vii. Paris, ca. 1900.

Pulfrich, C. "Ein neues Refractometer, besonders zum Gebrauch für Chemiker eingerichtet." *Zeitschrift für Instrumentenkunde* 8 (1888): 47–53.

Pulfrich, C. "Ein neues Refraktometer, Universalapparat für refraktometrische und spektrometrische Untersuchungen." *Zeitschrift für Instrumentenkunde* 15 (1895): 389–94; and *Zeitschrift für physikalische Chemie* 18 (1895): 294–99.

Pulfrich, C. "Über das neue Eintauchrefraktometer des Firma Carl Zeiss." *Zeitschrift für angewandte Chemie* (1899): 1168–70.

Pulfrich, C. *Das Totalreflectometer und das Refractometer für Chemiker.* Leipzig, 1890.

Schwarz, Robert. "The Use of the Immersion Refractometer in Examining American Beers Made from Malt and Unmalted Cereals." *Journal of Industrial and Engineering Chemistry* 5 (1913): 660–63.

Scientific Materials Co. *General Apparatus Catalogue.* Pittsburgh, 1915.

Simeon, F. "The Accuracy Attainable with Critical Angle Refractometers." *Proceedings of the Physical Society* 30 (1918); and *Chemical Abstracts* 13 (1919): 1971.

Skalweit, J. "On Butter Testing." *Analyst* 11 (1886): 90.

Spencer Lens Co. *Catalog of Spencer Products.* Buffalo, 1924.

Stanley, F. "Improved Types of British Refractometers." *Journal of the Society of Chemical Industry* 38 (1919): 142–43; and *Chemical Abstracts* 13 (1919): 2619.

Thomas, Arthur H. *Laboratory Apparatus and Reagents.* Philadelphia, 1921.

Thorpe, E., ed. *A Dictionary of Applied Chemistry*, vol 4. London, 1913.

Tolman, L. M., and L. S. Munson. "Refractive Indices of Salad Oils—Correction for Temperature." *Journal of the American Chemical Society* 24 (1902): 754–58.

Tolman, L. M., and W. B. Smith. "Estimation of Sugars by Means of the Refractometer." *Journal of the American Chemical Society* 28 (1906): 1476–82.

Twyman, F., and F. Simeon. "Accuracy Control in the Manufacture of Abbé and Pulfrich Refractometers." *Journal of the Society of Chemical Industry* 38 (1919): 142–243.

Wagner, B. *Tabellen zum Eintauchrefraktometer.* Sondershausen, 1907.

Washburn, Edward W., ed. *International Critical Tables of Numerical Data, Physics, Chemistry and Technology.* New York, 1926.

Wiley, Harvey W. "Dairy Products." U.S. Department of Agriculture, Division of Chemistry, *Bulletin* 13 (1887): 53.

Wiley, Harvey W. "Lard and Lard Adulterations." U.S. Department of Agriculture, Division of Chemistry, *Bulletin* 13 (1889): 441–43.

Wiley, Harvey W., ed. "Methods of Analysis Adopted by the Association of Official Agricultural Chemists." U.S. Department of Agriculture, Division of Chemistry, *Bulletin* 46 (1895): 31–32.

Wiley, Harvey W. *Principles and Practice of Agricultural Analysis.* Easton, PA: 1897.

Wiley, Harvey W. "Report of the Chemist." *Report of the Commissioner of Agriculture for 1887* (1888): 182.

Zeiss, Carl, Optical Works. *Abbé's Refractometers*, 3rd. ed. Jena, 1907.

Zeiss, Carl, Optical Works. *Das Butterrefraktometer*, 4th ed. Jena, 1907.

Zeiss, Carl, Optical Works. *Dipping Refractometer.* Jena, 1907.

Zeiss, Carl, Optical Works. *Ein newes Refraktometer für die Zuckerindustrie.* Jena, 1913.

Zeiss, Carl, Optical Works. *Industrial Refractometer.* Jena, 1926.

Zeiss, Carl, Optical Works. *Microscopes and Microscopical Accessories.* Jena, 1885.

Zeiss, Carl, Optical Works. *Optical Measuring Instruments.* Jena, 1893.

Zeiss, Carl, Optical Works. *Refraktometer Abbéscher Konstrucktion.* Jena, 1904.

Zeiss, Carl, Optical Works. *Tabelle zum Zeiss Butterrefraktometer.* (n.d.).

CHAPTER 4

Regional Styles in Pesticide Analysis

COULSON, LOVELOCK, AND THE DETECTION OF ORGANOCHLORINE INSECTICIDES

Peter J. T. Morris

Introduction

In this chapter, I examine two concurrent but quite different technological solutions to a key problem in environmental science: how to measure accurately and rapidly low levels of chlorinated hydrocarbon insecticides on fruit and vegetables. The use of one of these solutions (microcoulometry) has been restricted largely to the United States, while the other (the electron capture detector) has been more popular in Britain. This is all the more striking as the electron capture eventually became much more sensitive than the final version of the microcoulometric method, the Hall® electrolytic conductivity detector. However, this chapter is more than just a comparative account of two different ways of solving the same problem, it also uses the development of several new analytical methods—the Mills test for instance—to illustrate the growing concern of analytical chemists and regulatory bodies (but not at that stage, the public) about the increasing use of pesticides and the residues they left in vegetables, fruits, and dairy products. This concern had reached the American dairy industry by the late 1950s. This concern thus clearly predates the publication of Rachel Carson's *Silent Spring* in 1962 and it thus challenges a widely held perception of the impact of pesticide use was not taken seriously by industry in the 1950s.

Organochlorine Insecticides

The years immediately after the World War II saw the development of numerous new pesticides, notably the chlorinated hydrocarbons, following the introduction of DDT by the Swiss chemical firm Geigy in 1942 (see table 4.1).[1]

Since they were very effective at killing insects, had a low mammalian toxicity, and were comparatively cheap, DDT and its fellow organochlorine insecticides were soon used on a large scale. For instance, DDT production in the United States more than tripled from seventeen tonnes in 1953 to fifty-six tonnes in 1959.[2]

The standard method of detecting DDT on foodstuffs in the 1950s was the Schechter-Haller method introduced in 1945. DDT was nitrated and then treated with sodium methoxide to produce a colored derivate, the concentration of which was estimated using a colorimeter, an instrument that had been used in laboratories since the 1860s. Jules Duboscq introduced a comparative colorimeter in 1854; a later variant was the Klett colorimeter. J. W. Lovibond introduced his "Tintometer" in 1886; it is still used today in a practically unaltered form.[3] The Schechter-Haller method routinely detected DDT at 2.5 ppm, 1 ppm at best, and was also erratic. This lack of sensitivity was in contrast to the analysis of the older inorganic pesticides, such as white arsenic, that could be measured accurately to 0.2 parts per million.[4] These new insecticides were relatively inert chemically and thus were

TABLE 4.1

Insecticides Introduced between 1942 and 1951[a]

Insecticide	Introduced by	When
DDT	Geigy (Switzerland)	1942
lindane (gamma-BHC)	ICI (UK)	1945
chlordane	Velsicol (USA)	1945
DDD (TDE)	Rohm & Hass (U.S.)	1945
dieldrin	Julius Hyman[b] (U.S.)	1948
aldrin	Julius Hyman	1948
heptachlor	Velsicol	1948
endrin	Julius Hyman	1951

[a]Assembled from Hubert Martin (ed.), *Pesticide Manual: Basic Information on the Chemicals Used as Active Components of Pesticides*, 1st ed. (Worcester: British Crop Protection Council, 1968), supplemented by information from Erik Verg, Gottfried Plumpe, and Heinz Schultheis, *Milestones* (Leverkusen: Bayer AG, 1988).

[b]Julius Hyman was originally an employee of Velsicol, but he began the manufacture of chlordane as Julius Hyman and Company at a former mustard gas plant at the Rocky Mountain Arsenal, near Denver, in 1947. He developed the process for manufacturing dieldrin, aldrin, and endrin. He then lost a patent suit brought by Velsicol, and in 1952 his company was taken over by Shell, which already marketed his pesticides in California and neighboring states. The firm was completely absorbed into the Shell Chemical Corporation in 1954 and Hyman set up a private research laboratory in San Francisco. See www.pmrma.army.mil/site/s-plants.html (accessed July 18, 2007).

very different from the alkaloids that had been the staple of the poisoner and the toxicologist.[5] Furthermore, in contrast to the earlier inorganic pesticides, they were soluble in plant fats and waxes, thereby increasing the risk of food contamination.[6]

When the monitoring of marine wildlife began in the late 1950s, biologists (rather than chemists) were able to show that pesticides such as DDT and dieldrin had harmful effects on shrimps and crawfish at levels below existing detection limits.[7] It is striking that most of the evidence presented by Rachel Carson about the dangers of pesticides for marine life, her own research field, was based on biological experiments and not chemical analysis. The presence of the pesticides at such low levels in the late 1950s could only be inferred indirectly.

At the same time, food analysts and entomologists were urgently seeking methods of checking pesticide residues on fruits and vegetables that were rapid and reliable, but also cheap given the number of analyses required. The aim was to ensure that the crops met the tolerance requirements of the U.S. Food and Drug Administration, a process known as "screening" or "sorting." The tolerance limit was the maximum level accepted by the FDA in foodstuffs, basically the amount that was considered "tolerable" by the human body.

Analysts attempted to improve the Schechter-Haller method by using the relatively new routine ultraviolet spectrophotometers[8] to accurately measure the absorption at 596 nanometers.[9] Although it was better than the original colorimetric version of the Schechter-Haller method, the accuracy of the spectrophotometric method was not high. A 1961 paper[10] by Bill Durham of the U.S. Public Health Service on the DDT content of the body fat of Alaskan natives (one of the few analytical papers cited by Rachel Carson in *Silent Spring*[11]) had a typical detection limit of about 0.1 ppm at best. This method was also slow, required large amounts of the sample, was not specific for one pesticide, and could not be used for more than one pesticide at a time.[12] At a time when concern was growing about synergetic effect of mixtures of pesticides,[13] this latter failing was a major drawback. This was all against a background of a growing demand for analysts to detect insecticide residues—according to Francis Gunther, the doyen of American food analysts in the 1950s and 1960s who was based at the University of California, Riverside—at a level of 0.02 ppm in 1955 and 1 ppb five years later.[14]

Paul Mills of the U.S. Food and Drug Administration introduced a rapid semi-quantitative paper chromatography test for the detection of DDT residues in 1959.[15] It was sensitive to 0.02 ppm, but in practice the lower limit of detection was around 0.2 ppm. In the hands of a competent analyst, the Mills test worked well, but it had to be carried out according to the strict letter of the procedure to yield reliable results. It was thus highly sensitive to operator competence, particularly when the residue levels fell below 2 ppm.[16]

The pressure on food analysts increased by the passing in 1958 by the U.S. Congress of the Delaney Clause of the Food Additives Amendment to the 1938 Food, Drug and Cosmetic Act, which set a zero tolerance level in food for any substance found to be carcinogenic in animal or man. There were an increasing number of incidents involving the monitoring of foodstuffs. The most famous case was the cranberry scare of 1959, just before Thanksgiving in late November when cranberry consumption reached its annual peak. Although the U.S. Department of Agriculture had approved aminotriazole weed killer for use in cranberry bogs in 1957, subsequent trials showed that it might be carcinogenic. On November 9, Arthur

Flemming, then U.S. Secretary of Health, Education, and Welfare, announced that the FDA was going to check the entire cranberry crop for aminotriazole contamination. Although the FDA was eventually able to clear the crop before Thanksgiving, the sales of cranberries (unsurprisingly) fell sharply. In an attempt to avert a crisis, Vice President Richard Nixon consumed cranberries in public to show that they were safe to eat.[17] There does not appear to have been any similar scare involving DDT, nor were there any in Britain.

By the beginning of the 1960s, food and crops analysts on both sides of the Atlantic were becoming seriously concerned about their inability to detect both reliably and consistently very low levels of DDT and other chlorinated hydrocarbons. So much so, that the Pesticides in Foodstuffs Subcommittee of the Society for Analytical Chemistry in 1960 commissioned P. H. Needham of the Rothamsted Experimental Station to investigate bioassays as a means of rapidly testing for pesticide residues in foodstuffs.[18] Needham visited various laboratories that were using bioassay methods in the UK and in Europe. The sensitivity of the method was high. The city council laboratory in Zurich could detect DDT to a limit ranging from 0.1 to 0.005 ppm of using the larva of the yellow fever mosquito (*Aedes ægypti*). Clearly the method was useful but there were problems. Guinea pigs were needed to feed the mosquitoes and Needham recommended the fruit fly (*Drosophila melanogaster*) as the best insect to use. There were also variations between different foodstuffs; a level of pesticide that was toxic to mosquitoes in processed peaches was nontoxic in processed peas.

Shipment of evaporated milk and butter to Hawaii was impounded in early 1960 on the grounds that it was contaminated with pesticides. As a result of this latest scare, the (U.S.) Dairy Industry Committee—composed of the eight secretaries of dairy trade associations—appointed a technical advisory committee chaired by Henry E. O. Heineman of the Pet Milk Company, a major producer of evaporated milk, to look into the whole question. A working party of this committee carried out 31,548 tests on twenty-four milk products over two years, mainly using the Mills paper chromatography test.[19]

However, while these reports were being compiled, the instrumental situation was changing dramatically. The key element of a more accurate and rapid chemical analysis of pesticide was the gas-liquid chromatograph invented by Archer Martin and Tony James at the National institute for Medical Research (NIMR) in north London in the summer of 1951. This was arguably the most important advance in chemical analysis since Bunsen and Kirchhoff developed spectral analysis almost a century earlier.[20] However, for it to be useful, there had to be a reliable and sensitive method of identifying the different fractions as they emerged from the column. As late as 1956, Tony James was critical of existing detectors and concluded: "The field is open for the development of more sensitive and simpler detectors, and it is to be hoped that workers in this field will not rest content with the instruments at present available."[21]

Dale Coulson and Microcoulometry

By the mid-1950s, two successful analytical methods had been developed by the pesticide-manufacturing industry as existing methods were clearly unsatisfactory. One combined liquid-solid partition chromatography with combustion in a quartz tube followed by amperometric titration to determine the chlorine content.[22] The other used phenyl azide and dinitroaniline to form a colored compound with aldrin that was determined photometrically.[23] This method could be used with dieldrin if it was reduced before treatment with the phenyl azide.[24] Subsequently, the first practicable method of measuring organochlorine pesticide residues using gas chromatography was introduced by Dale Coulson of the Stanford Research Institute, Menlo Park, California, in 1959.[25] Coulson had taken his PhD at UCLA and had worked briefly for the Shell Development Company in 1951–1952 before he took up a position at the University of Colorado at Boulder in 1952–1953. His method combined the gas chromatograph with existing technique of combustion analysis. Coulson adopted the quartz tube furnace constructed for pesticide analysis by E. D. Peters at the Shell Development Company.[26] The chloride ions formed in this way were then analyzed using an automatic chloride analyzer that had been developed for this method by Coulson and his co-worker Leonard Cavanagh.[27] This microcoulometric titration method operated on the basis of a constant silver concentration in the titration cell and was partly based on earlier work by the eminent electrochemist James Lingane at Harvard and by O. E. Sundberg and his group at American Cyanamid, Bound Brook, New Jersey.[28] In effect he replaced the amperometric method used by Shell by a microcoulometric method developed by American Cyanamid. Coulometric titration of halides, specifically chlorides, was growing in popularity in this period because it "involves a minimum of manipulation. A further advantage is that the titrant is generated in situ in amounts that is [sic] measured in terms of physical quantities."[29] With high accuracy, excellent sensitivity, and relatively simple to operate, coulometric titration was ideal for the determination of "dilute solutions [of chlorides] with high precision" whether the aim was the determination of the atomic weight of chlorine or the amount of DDT on oranges.

In 1962, Jerry Burke and Loren Johnson of the U.S. Food and Drug Administration investigated[30] the analysis of seventy-one pesticides using the microcoulometric method, using a microcoulometric gas chromatograph marketed by Dohrmann Instruments and introduced—according to Gunther—in 1958.[31] They found that a cleanup procedure was necessary for blended vegetable samples, as natural constituents of vegetables interfered with the analysis and the recovery rates were often low. Nonetheless, they concluded, "When the limitations of the instrument are understood it should be a very useful tool for pesticide residue analysis."

Coulson was well-known in chemical circles in California—a state with a major fruit-growing industry and the home of the Montrose Chemical Corporation, the major American DDT manufacturer—in the early 1960s. His microcoulometric method was invented at Stanford Research Institute, the premier research institute in California. California was the leading state for the analysis of pesticide residues, particularly at the University of California's Citrus Research Station at Riverside. Coulson was part of this network and the wider community of agricultural chemists. His papers were published in *Agricultural and Food Chemistry* and he was awarded the Harvey W. Wiley Award of the Association of Official Agricultural Chemists[32] in 1974.[33] The Dohrmann Instrument version of his apparatus was described by fellow Californian chemist Francis Gunther in 1966 as "the outstanding example of a complete instrument specifically tailored to the requirements of pesticide-residue investigations."[34] It was also an advantage in California that the Coulson method worked well with citrus fruit, although it had problems analyzing the residues from some vegetables.

In 1965, Coulson introduced his own commercial model, but both the model and his company appear to have been short-lived.[35] On the basis of Coulson's work, Randall Hall of Purdue University developed a very sensitive microcoulometric detector[36] in 1974 that was smaller and easier to use. This has become known as the Hall electrolytic conductivity detector (confusingly called an HECD although it has nothing to do with electron capture detector; see below). The Hall detector was never popular in the UK, but it is still approved for use by the Environmental Protection Agency (EPA) in the United States.

James Lovelock and the Electron Capture Detector

Coulson's detector sprang from an existing technology of organochlorine and chloride analysis. In that sense, it was an incremental development. Its rival was a radical innovation[37] created in a wholly different environment and initially without any intention of analyzing organohalogen compounds.

In the early 1950s, James Lovelock, a medical researcher at the NIMR with a chemistry degree,[38] was researching the damage done to animal cells by freezing. He soon realized that it was connected to the fatty acid composition of the lipids in the cell membrane. How could these acids be measured in the minute concentrations found in laboratory animals? He turned to his colleagues Martin and James for help. They suggested that Lovelock used their new gas-liquid chromatograph, but unfortunately the titration method they used to detect the acids coming out of the column was too insensitive for the small amounts of fatty acids that Lovelock was studying.

Though Lovelock has claimed that the spur to develop a new detector was his desire to spend his time inventing rather than carrying out experiments to collect larger amounts of fatty acids, he was doubtlessly aware of the potential value of a more sensitive detector, especially for biochemical work. At this point, he recalled a sensitive anemometer he had constructed when he was working at the Common Cold Unit in Salisbury in the late 1940s. Made from radium scraped from old aircraft gauges, it worked on the principle of ion drift; the slow positive ions were easily disturbed by small air currents. Unfortunately, field trials in the Arctic had showed that it was also very sensitive to cigarette smoke! Perhaps if it were suitably modified it could be used as a detector.

In the early 1950s, John Otvos and David Stevenson at Shell Development had developed a beta-ray ionization chamber based on strontium-90 to analyze gases. The Emeryville team and their Dutch colleague Hendrik Boer soon realized the value of this device for gas chromatography; they developed their quite different detectors by 1955.[39] The Boer detector (now called a cross-section detector) was a dual chamber apparatus in which the effluent gas from the chromatograph flowed through one cell and the pure carrier gas through the other. The difference between the two ionization current in the two chambers was converted into a potential and measured by an electrometer. The voltage applied to the ionization chambers was between 100 and 300 volts and the carrier gas was usually nitrogen or hydrogen.

Lovelock combined their work with his anemometer to create the first-generation electron capture detector, which had a single chamber, thus removing the need to have a parallel flow of the pure carrier gas.[40] There was, however, a problem with the carrier gas. Light carrier gases such as hydrogen and helium would have worked best, but helium was not readily available as it was too expensive and hydrogen was unsuitable for overnight operation at high temperatures (and prohibited at the NIMR by fire regulations), so nitrogen was the usual carrier gas. Lovelock was able to reproduce Boer's results, but in his hands it was much less sensitive than other contemporary detectors and of no use for the delicate work he had in mind. He then decided to experiment with a lower potential, of approximately thirty volts. This seemed promising and James gave him a test mixture of fatty acid esters to try out. When Lovelock carried out the separation running the detector at 100 volts, he obtained four small peaks. Surprisingly, when he lowered the potential of the detector to ten volts, he observed numerous peaks running off the scale. Lovelock was eager to demonstrate his new detector to Martin and James, but matters did not turn out as he expected:

> I thought that the search was over and we now had a truly sensitive detector. I asked James and Martin to come try it, which they did, bringing with them an allegedly pure sample of methyl caproate. I shall never forget the look of amazement on Tony [James's] face as peak after peak was drawn from a small sample of this substance. Worse, none of them had the retention time of methyl caproate or of any other fatty acid ester. We now know that what was seen were traces of electron absorbing impurities in the sample, but at that time it seemed to be a useless and wholly anomalous device.[41]

Later experiments revealed that the electron capture detector was extremely sensitive to the presence of tiny amounts of certain impurities. The presence of a tiny amount of carbon tetrachloride in the silicone seal, for instance, was sufficient to render the detector useless.

Frustrated with the electron capture detector, Lovelock turned to the investigation of other ionization processes. He attempted to use a potential in the detector chamber that was greater than the ionization potential of most organic compounds but much less than that of the carrier gas. One day, the NIMR's stores were out of nitrogen and Lovelock was asked if argon would do. Since argon has a similar ionization potential to nitrogen, he agreed to use argon and ran his experiment with the same chamber he had used for his prototype electron capture detector at a very high potential of 700 volts. This setup gave splendid results with fatty acid esters and Lovelock thought the problem was solved. Alas, when the argon ran out and was replaced with the usual nitrogen, the sensitivity sank back to its former mediocre level.

Further investigation revealed that the improved performance was a result of the Penning effect. In 1934, Frans Michel Penning at Philips in Eindhoven had discovered that a metastable state of argon formed under these conditions transferring its energy to another gas molecule on collision as long as the ionization potential of the other molecule is less than the energy level of the metastable argon.[42] The organic compound is ionized by the argon atom rather than by direct electron capture. The number of molecules ionized in this way is fairly low, but still sufficient to allow effective detection of the eluates as they pass out of the chromatograph.

Lovelock unveiled the argon detector at an informal meeting of chromatographers in Oxford in 1957, and it was rapidly commercialized by W. G. Pye and Company, who launched the argon chromatograph at the Second International Gas Chromatography Symposium in Amsterdam in May 1958. Given the lack of suitable detectors that were available in the late 1950s, there was an immediate surge of requests for an argon detector from laboratories around the world.

When Lovelock went to New York in the spring of 1958 to give a paper about the argon detector, he met Seymour "Sandy" Lipsky, who was working on the metabolism of fatty acids in human plasma at Yale Medical School.[43] They found they had shared interests in the analysis of fatty acid esters and in ionization detectors. Subsequently, Lipsky invited Lovelock to work at Yale for several months in the academic year 1958–1959. As soon as Lovelock arrived, the two scientists decided to combine the argon detector with Marcel Golay's novel capillary column. With this combination, they were able to separate methyl oleate from its *trans* isomer methyl elaidate. They then published a seminal paper in the *Journal of the American Chemical Society*, which alerted analytical chemists to the potential value of the electron capture detector.[44] By this time, several scientific-instrument manufacturers had introduced their versions of argon detectors to be used in conjunction with their gas chromatographs.

Using the argon detector sold by Shandon Instruments, R. Goulden and his colleagues at the Shell Research Centre in Sittingbourne, Kent, published the first account of the detection of chlorinated hydrocarbon pesticides in 1960.[45] Soon afterward, J. O. Watts and A. K. Klein at the U.S. Food and Drugs Administration developed their own electron capture detector based on the paper published in 1960 by Lovelock and Lipsey.[46] A. D. Moore at the U.S. Forest Service also independently conceived the idea of analyzing insecticide residues using a Barber-Colman Model A-4071A (radium 226) argon detector in 1961.[47] All three groups showed that the electron capture detector could detect DDT and other chlorinated hydrocarbon pesticides to at least 1 ppm and often down to 0.25 ppm. By 1963, Watts and Klein had reduced the detection limit for DDT to 0.1 ppm, a level comparable with conventional chemical methods with a device that was still in an early stage of development.[48]

Lovelock was also able to use his time at Yale to turn the electron capture detector into a useable device. After a meeting with Ken McAffee of Bell Telephone Laboratories, Lovelock realized that the problems with the original electron capture detector were caused by the low potential used. If he could replace this low potential with high potential pulses, many of these problems would be solved. Eventually, he refined the idea even further to create a constant population of electrons, thereby converting a temperamental device into a reliable, stable, and relatively inexpensive detector. As with the argon detector, the makers of gas chromatographs were quick to incorporate the electron capture detector into their instruments from the winter of 1961 onward. Lovelock joined forces with Al Zlatkis when they were both working at the University of Houston in 1961–1962 to form Ionics Research, which supplied electron capture detectors to several firms, notably Perkin-Elmer.

Just as the ECD was being developed—in the late 1950s—as a result of public controversy over nuclear fallout from H-bomb tests, the U.S. Atomic Energy Authority tightened up the use of radium.[49] Whereas researchers in the UK freely used radium mainly from old aircraft gauges (later replaced by nickel-63), their U.S. counterparts were forced to use weaker sources, mainly tritium foils. These low-level sources were rather unsatisfactory.[50] There were also regulatory barriers to the use of commercial ECDs and a license was required to use it.[51] This was a major reason for the survival of microcoulometry: it did not use radioactive sources.

It appears that it was an incident that occurred in the winter of 1963 that first brought the electron capture detector to the fore in environmental analysis.[52] An abnormal number of fish were dying in the lower reaches of the Mississippi, but there was no obvious cause. Biological experiments soon revealed that there was a toxin in the river mud and in the livers of the dead fish, but what was it? The gas chromatograph and the electron capture detector revealed very low concentrations of two chemicals, which were later shown by analytical chemists to be intermediates in the production of endrin. Eventually, the source of these chemicals was traced to a dump used by a chemical plant at Memphis, Tennessee, owned by Velsicol. The power of the electron capture detector was subsequently demonstrated somewhat less dramatically by a report by the UK Ministry of Agriculture Fisheries and Food. Its authors were able to tabulate DDT residues in butter and milk to a hitherto unachievable limit of about 0.01 ppm using gas chromatography.[53] The detector is not mentioned but it was almost certainly an ECD given the radical improvement in detection levels.

Relative Performance

Gunther pointed out in 1966 that even one part per billion was the equivalent of "establishing that the extreme dryness of a particularly good martini is due to the fact that only a single drop of vermouth was added to 125,000 [U.S. gallons] of quality gin."[54] He later described detecting parts-per-trillion as "a fairy tale,"[55] but it is a fairy tale that has become fact.

In a 1963 paper that compared the electron capture detector with other methods of detecting pesticides in milk, Lloyd Henderson of Foremost Dairies, San Francisco, noted that the Mills test "has been found to be the most suitable for a rapid qualitative and semi-quantitative screening test."[56] He reported good results with the microcoulometric method, which could measure DDT residues accurately around 0.3 ppm and concluded that it yielded "probably more accurate results than the paper test when the total residues are 1.0 ppm or less." By contrast, he felt that:

> The electron-capture gas chromatographic procedure at this stage of its development yielded lower total residues with fewer pesticides identified than did the other two chromatographic procedures. The procedure, however, may be developed into a rapid and accurate method for the detection and evaluation of pesticide residues in milk and dairy products.[57]

The situation was later tilted in favor of the electron capture detector by Lovelock's continuing improvement of his device, and the current sensitivity of the electron capture detector is stated to be less than 10 femtograms (1×10^{14} g) of lindane, which is about 10 parts per *quadrillion*. The latest version of the Hall detector is said to have a sensitivity of 5 picograms (5×10^{12} g) of heptachlor or around 5 parts per trillion.[58] In short, the chemical analysis of chlorinated hydrocarbon residues has improved by up to 100 millionfold in the last three decades, but the electron capture detector is now 200 times more accurate than its rival.

By the mid-1970s, the electron capture detector and microelectrolytic methods had solved the problem of analyzing DDT residues to the degree that this formerly important (not to say fraught) topic had disappeared from the pages of the *Journal of the Association of Official Agricultural Chemists* by the mid-1970s. The pesticide analysis crisis was over.

Conclusion

The urgent need to analyze very low levels of pesticide residues in the late 1950s brought about two very different methods of analysis. The microcoulometric approach developed by Dale Coulson in California was an incremental innovation developed within a strong American tradition of electrochemical analysis and stemmed from earlier work within industry. It was aimed at food and crops analysts working on America's West Coast. By contrast, the electron capture detector was a radical innovation that was the product of the background of long-established medical research in Britain, which was initially aimed at analyzing small amounts of biological chemicals. It was also a result of the time-honored tradition in Britain of making your own equipment, especially in the cash-strapped 1950s. Nonetheless, it drew on research carried out within industrial laboratories.

The electron capture detector was the major method of detecting pesticide residues in Europe from the 1960s. On the West Coast of the United States, microcoulometry remained popular although the ECD was used as well. This was partly the result of the Californian origins of the Coulson method and its value for the analysis of citrus fruit, but largely a result of different regulations governing radioactive sources, with the implicit support of EPA authorization for the Hall detector. Regardless of the method used, with the fortuitous assistance of the development of the gas chromatograph, in less than a decade analytical chemists had solved the problem of measuring very low levels of pesticide residues. This involved research at chemical firms such as American Cyanamid,[59] and it is therefore ironic that the movement against the use of pesticides such as DDT in America and Europe in the late 1960s owed much to the development of analytical techniques that showed that traces of these pesticides were present in the environment and our bodies. Elsewhere I have argued that this reaction was grossly excessive and was partly based on a failure by the public at large to understand the insignificance of extremely low levels of pesticides.[60] Nonetheless, the pesticide industry was attacked with a sword that was largely of its own making. Nicholas Rasmussen remarked in his commentary on my earlier paper and one by Anthony Travis on American Cyanamid: "One wonders how many of these instruments would have been developed by the chemical industry in the 1930s through 1950s if environmental regulations and litigation pressures such as emerged in the 1960s had been present."[61]

Notes

1. For the history of DDT, see Dunlap, *DDT: Scientists, Citizens, and Public Policy*; Mellanby, *The DDT Story*; Simon, "DDT"; and Russell, "The Strange Career of DDT, 770–96. Broader surveys are provided by Sheail, *Pesticides and Nature Conservation*; and Bosso, *Pesticides and Politics*. Adam Curtis has produced a brilliant documentary history of DDT, "Goodbye Mrs Ant," as part of his documentary series *Pandora's Box*. This program was transmitted by the BBC on July 2, 1992. I wish to thank Michael Beasley of the National Museum of Photography, Film and Television, Bradford, for his help in tracking down the title and transmission date.

2. Kohn, "Agriculture, Pesticides and the American Chemical Industry," 163.

3. Johnston, "Colorimeter," 123–25. Warner, "The Duboscq Colorimeter." Stock, "The Duboscq Colorimeter and Its Inventor."

4. The long-established Gutzeit method was very sensitive for arsenic. In food, the detection limit laid down by the AOAC in 1925 was 0.001mg, and as the initial sample (for vegetables) was 25 g, this equates to 1 ppm (as the original solution of the digested sample would be divided into four aliquots). By 1950, the AOAC laid down that the concentration should be recorded to not more than three decimal places as grains per pound. This implies a detection limit of around 0.2 ppm. See Doolittle, *Official and Tentative Methods of Analysis of the Association of Official Agricultural Chemists*, 171–73; and Lepper, *Official Methods of Analysis of the Association of Official Agricultural Chemists*, 369–73.

5. The development of organic toxicology is yet another unplowed field in the history of chemistry. For a possible starting point for such a history, see Watson, Wexler, and Everitt, "History," 1–5.

6. Gunther and Blinn, *Analysis of Insecticides and Acaricides*, 12.

7. Carson, *Silent Spring*, 124, citing a private communication from John C. Pearson. Also see Newson, "Some Ecological Implications of Two Decades of Use of Synthetic Organic Insecticides for Control of Agricultural Pests in Louisiana," 451 and table 24–7. I am indebted to George Twigg for the latter reference.

8. Eklund and Morris, "Spectrophotometer," 558–61.

9. For the technical details see Gunther and Blinn, *Analysis of Insecticides and Acaricides*, 412–15.

10. Durham, et al., "Insecticide Content of Diet and Body Fat of Alaskan Natives."

11. Carson, *Silent Spring*, 19.

12. Rosen and Gretch, "Analysis of Pesticides: Evolution and Impact," 129.

13. U.S. Food and Drug Administration, *Facts for Consumers*, 6.

14. Gunther, "Advances in Analytical Detection of Pesticides," 276–302, and table 2. It is not clear what Gunther meant by "analytical sensitivity (or minimum detectability) requirements," it does not appear to be the same as the tolerance limits set by the FDA or the limits of existing methods. Gunther, in 1955, had characteristically claimed that "claimed sensitivities of '0.1 p.p.m.' can have little realistic significance" (see Gunther and Blinn, *Analysis of Insecticides and Acaricides*, 120). However, these figures do illustrate the pressure pesticide analysts felt they were under in the late 1950s.

15. Mills, "Detection and Semiquantitative Estimation of Chlorinated Pesticides Residues in Foods by paper Chromatography."

16. Henderson, "Comparison of Laboratory Techniques for the Determination of Pesticide residues in Milk."

17. Dunlap, *DDT*, 107–8; Carson, *Silent Spring*, 183–84; Beatty, *The DDT Myth*, 153–55.

18. Needham, "An Investigation into the Use of Bioassay for Pesticide Residues in Foodstuffs."

19. Henderson, "Comparison of Laboratory Techniques," 210–11.

20. The history of gas chromatography is sadly undeveloped. The best available sources are Ettre, "Gas Chromatography," and Ettre and Zlatkis, *75 Years of Chromatography*. Also see Ophield, "Separated for Thirty Years."

21. Keulemans, *Gas Chromatography*, 99.

22. For the technical details see Gunther and Blinn, *Analysis of Insecticides and Acaricides*, 357–70.

23. See Gunther and Blinn, *Analysis of Insecticides and Acaricides*, 268–74.

24. For its application to dieldrin, also see Gunther and Blinn, *Analysis of Insecticides and Acaricides*, 431.

25. Coulson, Cavanagh, and Stuart, "Gas Chromatography of Pesticides"; and Coulson, Cavanagh, de Vries, and Walther, "Microcoulometric Gas Chromatography of Pesticides." Also see U.S. Patent 3,032,493 filed on December 31, 1959, by Dale Coulson and Leonard Cavanagh and assigned to the Dohrmann Instruments Co. of San Francisco.

26. Peters, Rounds, and Agazzi, "Determination of Sulfur and Halogens: Improved Quartz Tube Combustion Analysis"; and Agazzi, Peters, and Brooks, "Combustion Techniques for the Determination of Residues of Highly Chlorinated Pesticides."

27. Coulson and Cavanagh, "Automatic Chloride Analyzer."

28. Lingane, "Automatic Coulometric Titration with Electrolytically Generated Silver Ion: Determination of Chloride, Bromide and Iodide Ions"; and Sundberg, Craig, and Parsons, "Determination of Halogen in Organic Compounds by Automatic Coulometric Titration."

29. Marinenko and Taylor, "Precise Coulometric Titrations of Halides."

30. Burke and Johnson, "Investigations in the Use of the Micro-Coulometric Gas Chromatograph for Pesticide Residue Analysis."

31. Gunther, "Advances in Analytical Detection of Pesticides," 294. This date appears to be at odds with the filing of the key patent for Dohrmann a year later.

32. Later the Association of Official Analytical Chemists and now AOAC International.

33. His acceptance speech, a partial autobiography that tells us nothing about his work on microcoulometry, was published in the *Journal of the AOAC* 58 (1975): 174–83.

34. Gunther, "Advances in Analytical Detection of Pesticides," 294.

35. Coulson, "Electrolytic Conductivity Detector for Gas Chromatography." The Coulson Instrument Co. was founded in 1964, according to the Coulson entry in *American Men of Science*. However, I have not come across any instruments made by this company and cannot find any trade literature, or indeed any mention of it on the World Wide Web. Curiously, Coulson makes no reference to his company or even his chromatography in his Harvey Wiley Award speech in 1974, reproduced in *Journal of the AOAC* 58 (1975): 174–83. It appears that it may have been assumed into SRI Co., the commercial wing of the Stanford Research Institute, where Coulson worked.

36. Hall, "A Highly Sensitive and Selective Microelectrolytic Conductivity Detector for Gas Chromatography."

37. For a discussion of the differences between incremental and radical innovations, see Morris, *The American Synthetic Rubber Research Program* and the references given therein.

38. Lovelock's own account of the development of the electron capture detector and his career can be found in Lovelock, *Homage to Gaia*.

Also see Lovelock's essay in Ettre and Zlatkis, *75 Years of Chromatography*; his historical account in Zlatkis and Poole, *Electron Capture: Theory and Practice in Chromatography*; and his entry on the "Electron Capture Detector," in Bud and Warner, *Instruments of Science: An Historical Encyclopedia*.

39. Keulemans, *Gas Chromatography*, 80–81. Also see the essay by Boer in Ettre and Zlatkis, *75 Years of Chromatography*, 14–15.

40. For technical descriptions, see Lovelock, "Ionization Methods for the Analysis of Gases and Vapors"; Lovelock and Gregory, "Electron Capture Ionization Detectors," in *Gas Chromatography: Third International Symposium Held Under the Auspices of the Analysis Instrumentation Division of the Instrument Society of America, June 13–16, 1961*. Lovelock, "The Electron Capture Detector: Theory and Practice"; and Lovelock and Watson, "The Electron Capture Detector: Theory and Practice II."

41. Lovelock, *Homage to Gaia*, 195–96.

42. Penning, "The Starting Potential of the Glow Discharge in Neon Argon Mixtures Between Large Parallel Plates. II Discussion of the Ionisation and Excitation of Electrons and Metastable Atoms"; and Ruffner, *Eponyms Dictionaries Index: A Reference Guide to Persons, Both Real and Imaginary, and the Terms Derived From Their Names*.

43. See the essay by Lipsky in Ettre and Zlatkis, *75 Years of Chromatography*, 265–76.

44. Lovelock and Lipsky, "Electron Affinity Spectroscopy—A New Method for the Identification of Functional Groups in Chemical Compounds Separated by Gas Chromatography."

45. Goodwin, Goulden, Richardson, and Reynolds, "The Analysis of Crop Extracts for Traces of Chlorinated Pesticides by Gas-Liquid Partition Chromatography." Also see Goodwin, Goulden, and Reynolds, "Rapid Identification and Determination of Residues of Chlorinated Pesticides in Crops by Gas-Liquid Chromatography."

46. Watts and Klein, "Determination of Chlorinated Pesticide Residues by Electron-Capture Gas Chromatography"; Lovelock and Lipsky, "Electron Affinity Spectroscopy," 431–33.

47. Moore, "Electron Capture with an Argon Ionization Detector in Gas Chromatographic Analysis of Insecticides." In a footnote, Moore remarks, "After this manuscript was submitted it was brought to the authors [sic] attention that E. S. Goodwin, R. Goulden, A. Richardson and J. G. Reynolds have also used the Shandon argon detector as an electron capture detector for studies of insecticide residues." One wonders if Goulden was one of the referees.

48. Klein, Watts, and Damico, "Electron Capture Gas Chromatography for the Determination of DDT in Butter and Some Vegetable Oils."

49. Walker, *Permissible Dose: A History of Radiological Protection in the Twentieth Century*, 18–28.

50. See the essay by Lipsky in Ettre and Zlatkis, *75 Years of Chromatography*, 271.

51. Ettre, *Chapters in the Evolution of Chromatograpy*, p. 331.

52. Graham, *Since Silent Spring*, 96–108.

53. Advisory Committee on Poisonous Substances Used in Agriculture and Food Storage, *Review of the Persistent Organochlorine Pesticides*, 13 and appendix F.

54. Gunther, "Advances in Analytical Detection of Pesticides," 293.

55. Beatty, *DDT Myth*, 26.

56. Lloyd Henderson, "Comparison of Laboratory Techniques."

57. Ibid, 215.

58. Hall detector sensitivity, http://tmqaustin.com/Detectors/t2mdetspecs.htm (accessed 29 March 2000).

59. For an example of how chemical firms developed instruments and techniques for low-level concentrations of toxic and other contaminants see Travis, "Instrumentation in Environmental Analysis, 1935–1975."

60. Morris, "'Parts per Trillion is a Fairy Tale,'" 259–84.

61. Rasmussen, "Innovation in Chemical Separation and Detection Instruments: Reflections on the Role of Research-Technology in the History of Science," 254.

References

Advisory Committee on Poisonous Substances Used in Agriculture and Food Storage. *Review of the Persistent Organochlorine Pesticides*. London: HMSO, 1964.

Agazzi, E. J., E. D. Peters, and F. R. Brooks. "Combustion Techniques for the Determination of Residues of Highly Chlorinated Pesticides." *Analytical Chemistry* 25 (1953): 237–40.

American Men of Science, Physical and Biological Sciences, 11th ed., vol. A–C. New York: R. R. Bowker Co., 1965.

Andrewes, C. S. *In Pursuit of the Common Cold*. London: Heinemann Medical, 1973.

Anonymous. *South Plants Area*. Available at http://www.pmrma.army.mil/site/splants.html (accessed July 18, 2007).

Bosso, Christopher J. *Pesticides and Politics: The Life Cycle of a Public Issue*. Pittsburgh: University of Pittsburgh Press, 1987.

Burke, Jerry, and Loren Johnson. "Investigations in the Use of the Micro-Coulometric Gas Chromatograph for Pesticide Residue Analysis." *Journal of the AOAC* 45 (1962): 348–54.

Carson, Rachel. *Silent Spring*. London: Hamish Hamilton, 1963.

Coulson, Dale M. "Electrolytic Conductivity Detector for Gas Chromatography." *Journal of Gas Chromatography* 3 (April 1965): 134–37.

Coulson, Dale M. "Acceptance Speech." *Journal of the AOAC* 58 (1975): 174–83.

Coulson, Dale M., and Leonard A. Cavanagh. "Automatic Chloride Analyzer." *Analytical Chemistry* 32 (1960): 1245–47.

Coulson, Dale M., Leonard A. Cavanagh, and Janet Stuart. "Gas Chromatography of Pesticides." *Agricultural and Food Chemistry* 7 (1959): 250–51.

Coulson, Dale M., Leonard A. Cavanagh, John E. de Vries, and Barbara Walther. "Microcoulometric Gas Chromatography of Pesticides." *Agricultural and Food Chemistry* 8 (1960): 399–402.

Curtis, Adam. "Goodbye Mrs. Ant." *Pandora's Box*. BBC, July 2, 1992.

Doolittle, R. E. *Official and Tentative Methods of Analysis of the Association of Official Agricultural Chemists*. 2nd ed. Washington, DC: Association of Official Agricultural Chemists, 1925.

Dunlap, Thomas R. *DDT: Scientists, Citizens, and Public Policy*. Princeton: Princeton University Press, 1981.

Durham, William F., et al. "Insecticide Content of Diet and Body Fat of Alaskan Natives." *Science* 134 (1961): 1880–81.

Eklund, Jon, and Peter Morris. "Spectrophotometer." In *Instruments of Science: An Historical Encyclopedia*, edited by Robert Bud and D. J. Warner, 558–61. New York and London: Garland, 1998.

Ettre, Leslie S. "Gas Chromatography." In *A History of Analytical Chemistry*, edited by Herbert A. Laitinen and Galen W. Ewing, 296–306. Washington, DC: Division of Analytical Chemistry of the American Chemical Society.

Ettre, L. S., and A. Zlatkis, eds. *75 Years of Chromatography*. Amsterdam: Elsevier, 1979.

Goodwin, E. S., R. Goulden, A. Richardson, and J. G. Reynolds. "The Analysis of Crop Extracts for Traces of Chlorinated Pesticides by Gas-Liquid Partition Chromatography." *Chemistry and Industry* (September 24, 1960): 1220–21.

Goodwin, E. S., R. Goulden, and J. G. Reynolds. "Rapid Identification and Determination of Residues of Chlorinated Pesticides in Crops by Gas-Liquid Chromatography." *Analyst* 86 (1961): 697–709.

Graham, Frank, Jr. *Since Silent Spring*. London: Hamish Hamilton, 1970.

Gunther, Francis A. "Advances in Analytical Detection of Pesticides." In *Scientific Aspects of Pest Control*. Washington, DC: National Academy of Sciences/National Research Council, 1966, 276–302.

Gunther, Francis A., and Roger C. Blinn. *Analysis of Insecticides and Acaricides*. New York: Interscience Publishers, 1955.

Hall, Randall C. "A Highly Sensitive and Selective Microelectrolytic Conductivity Detector for Gas Chromatography." *Journal of Chromatographic Science* 12 (March 1974): 152–60.

Hall detector sensitivity, http://tmqaustin.com/Detectors/t2mdctspecs.htm (accessed 29 March 2000).

Henderson, Lloyd. "Comparison of Laboratory Techniques for the Determination of Pesticide Residues in Milk." *Journal of the AOAC* 46 (1963): 209–15.

Johnston, Sean. "Colorimeter." In *Instruments of Science: An Historical Encyclopedia*, edited by Robert Bud and D. J. Warner. New York and London: Garland, 1998, 123–25.

Keulemans, A. I. M. *Gas Chromatography*. 2nd ed. New York: Reinhold, 1959.

Klein, A. K., J. O. Watts, and J. A. Damico. "Electron Capture Gas Chromatography for the Determination of DDT in Butter and Some Vegetable Oils." *Journal of the AOAC* 46 (1963): 165–71.

Kohn, Gustave. "Agriculture, Pesticides and the American Chemical Industry." In *Silent Spring Revisited*, edited by Gino J. Marco, Robert M. Hollingsworth, and William Durham, 163. Washington, DC: American Chemical Society, 1987.

Lepper, Henry A. *Official Methods of Analysis of the Association of Official Agricultural Chemists*. 7th ed. Washington, DC: Association of Official Agricultural Chemists, 1950.

Lingane, James J. "Automatic Coulometric Titration with Electrolytically Generated Silver Ion: Determination of Chloride, Bromide and Iodide Ions." *Analytical Chemistry* 26 (1954): 622–26.

Lipsky, S. R. In *75 Years of Chromatography*, edited by L. S. Ettre and A. Zlatkis, eds., 271. Amsterdam: Elsevier, 1979.

Lovelock, James E. *Homage to Gaia*. Oxford: Oxford University Press, 2000.

Lovelock, James E. In *75 Years of Chromatography*, edited by L. S. Ettre and A. Zlatkis, eds., 277–84. Amsterdam: Elsevier, 1979.

Lovelock, James. "The Electron Capture Detector." In *Instruments of Science: An Historical Encyclopedia*, edited by Robert Bud and Deborah Jean Warner, 213–14. New York and London: Garland, 1998.

Lovelock, James. "The Electron-Capture Detector—A Personal Odyssey." In *Electron Capture: Theory and Practice in Chromatography*. Journal of Chromatography Library, vol. 20, edited by A. Zlatkis and C. F. Poole. New York: Elsevier, 1981: 1–11.

Lovelock, J. E. "Ionization Methods for the Analysis of Gases and Vapors." *Analytical Chemistry* 33 (1961): 162–78.

Lovelock, J. E. "The Electron Capture Detector: Theory and Practice." *Journal of Chromatography* 99 (1974): 3–12.

Lovelock, J. E., and A. J. Watson. "The Electron Capture Detector: Theory and Practice II." *Journal of Chromatography* 158 (1978): 123–38.

Lovelock, J. E., and N. L. Gregory. "Electron Capture Ionization Detectors." In *Gas Chromatography: Third International Symposium Held Under the Auspices of the Analysis Instrumentation Division of the Instrument Society of America, June 13–16, 1961*, edited by Nathaniel Brenner, Joseph E. Callen, and Marvin D. Weiss. New York and London: Academic Press, 1961: 219–29.

Lovelock, J. E., and S. R. Lipsky. "Electron Affinity Spectroscopy—A New Method for the Identification of Functional Groups in Chemical Compounds Separated by Gas Chromatography." *Journal of the American Chemical Society* 82 (1960): 431–33.

Marinenko, George, and John K Taylor. "Precise Coulometric Titrations of Halides." *Journal of Research of the National Bureau of Standards—A. Physics and Chemistry* 67A (January–February 1963): 31–35. http://nvl.nist.gov/pub/nistpubs/jres/067/1/V67.N01.A05 (accessed July 31, 2007).

Martin, Hubert, ed. *Pesticide Manual: Basic Information on the Chemicals Used as Active Components of Pesticides*, 1st ed. Worcester: British Crop Protection Council, 1968.

Mellanby, Kenneth. *The DDT Story*. Farnham: British Crop Protection Council, 1992.

Mills, Paul A. "Detection and Semiquantitative Estimation of Chlorinated Pesticides Residues in Foods by Paper Chromatography." *Journal of the AOAC* 42 (1959): 734–40.

Moore, A. D. "Electron Capture with an Argon Ionization Detector in Gas Chromatographic Analysis of Insecticides." *Journal of Economic Entomology* 55 (1962): 271–72.

Morris, Peter J. T. "'Parts per Trillion is a Fairy Tale': The Development of the Electron Capture Detector and its Impact on the Monitoring of DDT." In *From Classical to Modern Chemistry: The Instrumental Revolution*, edited by Peter J. T. Morris. Cambridge: Royal Society of Chemistry in association with the Science Museum, 2002, 259–84.

Morris, Peter J. T. *The American Synthetic Rubber Research Program*. Philadelphia: University of Pennsylvania Press, 1989.

Needham, P. H. "An Investigation into the Use of Bioassay for Pesticide Residues in Foodstuffs." *Analyst* 85 (1960): 792–809.

Newson, L. D. "Some Ecological Implications of Two Decades of Use of Synthetic Organic Insecticides for Control of Agricultural Pests in Louisiana." In *The Careless Technology: Ecology and International Development*, edited by M. Taghi Farvar and John P. Milton, 451. Garden City, NY: Natural History Press, 1972.

Ophield, Mike. "Separated for Thirty Years." *Laboratory Practice* 37/8 (1988): 19–23.

Penning, F. M. "The Starting Potential of the Glow Discharge in Neon Argon Mixtures Between Large Parallel Plates II. Discussion of the Ionisation and Excitation of Electrons and Metastable Atoms." *Physica* 1 (1934): 1028–44.

Peters, E. D., G. C. Rounds, and E. J. Agazzi. "Determination of Sulfur and Halogens: Improved Quartz Tube Combustion Analysis." *Analytical Chemistry* 24 (1952): 710–14.

Rasmussen, Nicholas. "Innovation in Chemical Separation and Detection Instruments: Reflections on the Role of Research-Technology in the History of Science." In *From Classical to Modern Chemistry: The Instrumental Revolution*, edited by Peter J. T. Morris. Cambridge: Royal Society of Chemistry in association with the Science Museum, 2002, 254.

Rosen, Joseph D., and Fred M. Gretch. "Analysis of Pesticides: Evolution and Impact." In *Silent Spring Revisited*, edited by Gino J. Marco, Robert M. Hollingsworth, and William Durham, 129. Washington, DC: American Chemical Society, 1987.

Ruffner, James A., ed. *Eponyms Dictionaries Index: A Reference Guide to Persons, Both Real and Imaginary, and the Terms Derived From Their Names*. Detroit: Gale Research Co., 1977.

Russell, Edmund P. III. "The Strange Career of DDT: Experts, Federal Capacity and Environmentalism in World War II." *Technology and Culture* 40 (1999): 770–96.

Sheail, John. *Pesticides and Nature Conservation: The British Experience, 1950–1975.* Oxford: Oxford University Press, 1985.

Simon, Christian. "DDT." *Kulturgeschichte einer chemischen Verbindung.* Basel: Christoph Merian Verlag, 1999.

Stock, John T. "The Duboscq Colorimeter and Its Inventor." *Journal of Chemical Education* 71 (November 1994): 967–70.

Sundberg, O. E., H. C. Craig, and J. S. Parsons. "Determination of Halogen in Organic Compounds by Automatic Coulometric Titration." *Analytical Chemistry* 30 (1958): 1842–46.

Travis, Anthony S. "Instrumentation in Environmental Analysis, 1935–1975." In *From Classical to Modern Chemistry: The Instrumental Revolution*, edited by Peter J. T. Morris, 285–308. Cambridge: Royal Society of Chemistry in association with the Science Museum, 2002.

U.S. Department of Agriculture, Community Stabilization Scheme. *The Pesticide Situation.* Washington, DC: Government Printing Office, annual, 1953–1960.

U.S. Food and Drug Administration. *Facts for Consumers: Pesticide Residues.* Washington, DC: Food & Drug Administration, 1963.

Verg, Erik, Gottfried Plumpe, and Heinz Schultheis. *Milestones.* Leverkusen: Bayer AG, 1988.

Walker, J. Samuel. *Permissible Dose: A History of Radiological Protection in the Twentieth Century.* Berkeley: University of California Press, 2000.

Warner, Deborah Jean. "The Duboscq Colorimeter." *Bulletin of the Scientific Instrument Society* 88 (2006): 68–70.

Watson, Katherine D., Philip Wexler, and Janet M. Everitt. "History." In *Information Resources in Toxicology*, 3rd ed., edited by Philip Wexler, Pertti J. Hakkinen, Gerald Kennedy, and Frederick W. Stoss, pp. 1–25. San Diego and London: Academic Press, 1999.

Watts, J. O., and A. K. Klein. "Determination of Chlorinated Pesticide Residues by Electron-Capture Gas Chromatography." *Journal of the AOAC* 45 (1962): 102–8.

Instruments in
the Museum

CHAPTER 5

Hard Times— The Difficult Lives of Three Instruments in the Museum

Christian Sichau

Introduction

On a very hot day in July 1922, the curator of physics at the Deutsches Museum, Franz Fuchs, gave a talk to the entire staff of the museum. Mechanics, joiners, sculptors from the various workshops as well as members of the administration had to listen to his "Suggestions for Exhibiting Relativity Theory."[1] According to his memoirs, the idea for such a presentation as a test for an exhibition had come from Oskar von Miller, the founder of the museum, who wanted relativity theory to be elucidated in the museum on a popular level. Fuchs had already prepared some charts on "the addition of velocities, the Michelson-Experiment, the simultaneousness etc" which he now tried to explain to this audience. The result was disappointing: "The charts were not produced."[2] Thus, with the opening of the new museum building coming closer, novel ideas how to present this fashionable and much discussed scientific theory were needed. Since Einstein had been a member of the advisory body of the museum since 1920, he was asked directly for advice in 1924. However, the museum got nothing out of him beyond a suggestion to invite Hans Thirring, professor of physics at Vienna, to participate in this project. Thirring had just published the book *The Idea of Relativity Theory*, which was aimed at a popular readership.[3] Fuchs specified the museum's wishes in his letter to Thirring: the main aspects and problems that led to the rise of relativity should be explained without any mathematics and the line of thought should be sketched on how Einstein's theory provided solutions for these problems; additionally, the meaning of these solutions for the natural sciences in general might be discussed. For all this, six square meters on a wall would be reserved. With some hesitation, Thirring recommended to make use of a table shown on the last page of his book, summarizing in an abstract way the "genealogical tree" of relativity theory (see figure 5.1). He pointed out, however, that the table might be difficult to understand and therefore not of much use, but in order to make it comprehensible the whole text of the book would be needed.[4] This chart was also not produced. The plans to exhibit relativity theory were dropped. Only some space on a wall was reserved for a portrait of Einstein—but according to Fuchs' memoirs even this idea did not materialize. Relativity was an abstract, a bookish theory and it seemed impossible to give an account of it within a museum.

But does relativity really exist only in the realm of ideas and of mathematics? Is there no relativity to be found in a laboratory or workshop and thus in material objects that might be displayed at a museum? To what extent is our perception of relativity only a result of a specific historical process in which various prominent actors—among them Einstein himself—tried (successfully) to portray relativity as a result of abstract thought and thus as an exemplar of the new subdiscipline "theoretical physics," struggling for its legitimization?[5]

This episode from the beginning of the twentieth century serves to highlight some of the problems we are confronted with in the long history of exhibiting science and its history. At least on a theoretical level we have learned a lot, and a thorough discussion has been going on in recent years of presenting science not as a set of ideas but as a process in which

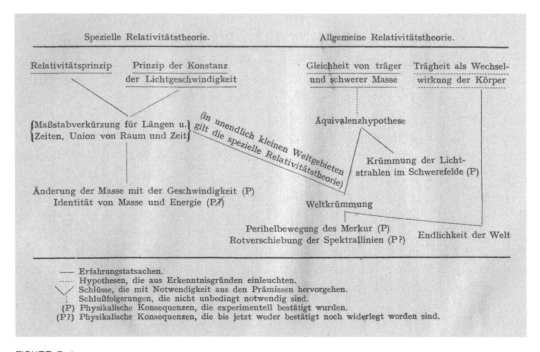

FIGURE 5.1
The "genealogical tree" ("*der Stammbaum*") of Relativity according to H. Thirring. Courtesy Deutsches Museum, Munich.

material objects play a key role.[6] But have we really been successful? The traditional views as expressed by Fuchs are still current today. Given the strong influence of science education (in a very narrow interpretation), scientific theories, concepts, or ideas and their development still dominate many exhibitions; instruments and experimental setups remain mere illustrations of abstract theories, of "ideas" that otherwise could not be displayed in museums.[7] Instruments and experimental setups are not treated as real objects possessing their own identity and history, histories that quite often are different from those traditionally told. Instead of asking how we can make these objects fit the widespread simplified big pictures of science and its development or a prefabricated storyline of an exhibition, we might be much better off if we try to build up our stories from the "biographies" of these individual objects.[8] Otherwise, opportunities are missed, as I will now try to show by following the path of some instruments from the laboratory to the museum and by analyzing their changing meaning during this journey.

Misplaced Objects, Missed Opportunities

In order to address this point we will go back to the year 1922 and the Deutsches Museum. When Franz Fuchs tried to find a way to exhibit relativity in the physics galleries, he himself was at the very same time organizing another new gallery on "electrical rays." He wrote letters to various institutions and scientists asking for instruments that could be shown at the museum. Since there existed already a strict conceptual outline for the planned exhibition every object had been appointed its "proper" place.[9] Fuchs had to face the usual difficulties: Instruments were in some cases still used and needed in the laboratories of the scientists, or the instruments had been reused and changed for different purposes; or they were simply destroyed or lost. From Walter Kaufmann, professor of physics at the University of Königsberg, the museum did not get exactly what it wanted. Kaufmann sent only an apparatus that he had used to measure the charge to mass ratio of very rapid electrons, since all of his other instruments had been taken apart, destroyed, or lost (see figure 5.2).[10] Alfred Bucherer, professor of physics at the University of Bonn, offered a similar apparatus with which he had worked in 1908 (see figure 5.3).[11] All these instruments fitted (seemingly) nicely in the "Entwicklungsreihe," the genealogy, of experiments on the charge of the electron. They had their appropriate place.

However, this is only true within the given overall outline of the physics galleries in the Deutsches Museum and it only holds as long as we do not take any closer look into the histories of these objects. For a start, a phrase Kaufmann used in his correspondence with the museum about his apparatus reveals that he thought about it somewhat differently. He pointed toward the aim of the experiments, which was "to determine the *change of mass of rapid electrons*" (my emphasis). So let us look more closely on what had originally been done with the help of these two instruments.

Kaufmann's experimental research on the electron lasted for several years; the instrument given to the museum reflects therefore of course only a small part of it. A description of it had been given by Kaufmann in a publication of the year 1906 titled "On the Constitution

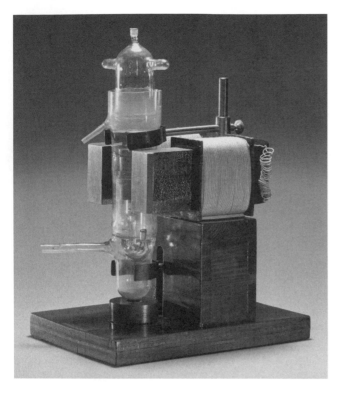

FIGURE 5.2
The apparatus by Walter Kaufmann for measuring the change of mass of very rapid electrons, 1906. Coutesy Deutsches Museum, Munich.

of the Electron" (see figure 5.4).[12] Fast electrons (obtained from radioactive β-decay) were deflected by a combination of a magnetic and an electrostatic field and the amount of the deflection was registered with a small photographic plate. Due to their variation in velocity and depending on the applied fields each point on the photographic plate corresponded to an electron with a specific velocity and mass.[13] The central part of the instrument was a small metal cylinder, about 55 millimeters in height and 32 millimeters in diameter. The photographic plate is at the top, the radioactive source at the bottom of the cylinder. In the lower half you will find the plate capacitor. In identical form the sketch of the instrument appeared also one year later in a different article written by Albert Einstein. It was his first lengthy discussion of "The Principle of Relativity and its Consequences."[14] Kaufmann's experiments were described in detail by Einstein because, as he wrote, the electrons move so fast in this apparatus that they should show relativistic effects. There was a second reason for Einstein to discuss these experiments carefully: Kaufmann's result did *not confirm* Einstein's prediction. Now, this leads directly to our second apparatus, the one by Bucherer. He presented his work in 1909 with the title "The Experimental Confirmation of the Principle of Relativity" and evoked a renewal of the discussion about Kaufmann's experiments.[15] Bucherer had made some changes to the configuration of the experiment, the radioactive source was in the center between two circular plates and the radiation was recorded on a photographic film fixed to the wall of the metal cylinder.

FIGURE 5.3
The apparatus by Alfred Bucherer for measuring the change of mass of very rapid electrons, 1908. Courtesy Deutsches Museum, Munich.

At least in Bonn, where Kaufmann and Bucherer had both worked, relativity had entered the laboratory and theory had encountered experiment. Without going further into details of this story, we can clearly see that both experiments, Kaufmann's and Bucherer's, had played a very important role in the early history of the special theory of relativity.[16] This episode also indicates a shift in the discussion and presentation of relativity: In 1907 Einstein had analyzed Kaufmann's results at length and took part in a long and intensive discussion among physicists about the experimental difficulties involved; nevertheless, in 1922 in Fuchs's talk to museum personnel neither of these experiments was mentioned at all. This experimental dimension of relativity was forgotten, and thus, the two instruments were not included in the planned exhibition on the theory of relativity. Instead, relativity had become an abstract theory, outside the world of laboratories. Here, the museum followed largely the widespread view in the wider public. It did not offer a different standpoint that might have been developed on the basis of the instruments at hand. Further, to incorporate these two instruments into the gallery on "electrical rays" meant to be true to the very aims of museums: to put the historical development of science into a strict order based on the state of *present* scientific knowledge. By doing so the progress of science should be made evident and complex instruments supposedly more readily comprehensible. Instruments were thus exhibited in order to show what was known about the world at that time and not to illustrate within which conceptual frameworks knowledge about the world was sought in the past and how this has shaped the research in the present.

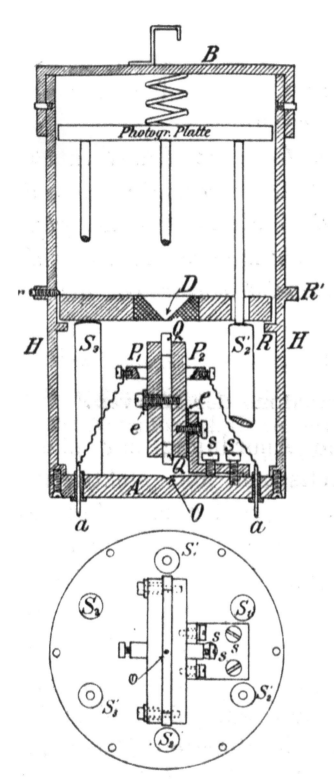

FIGURE 5.4
Detailed sketch of the inner cylinder of
Kaufmann's apparatus. *O* marks the
radioactive source for the electrons; PP is
the plate capacitor; the photographic
plate is at the top. From Kaufmann, "Über
die Konstitution des Elektrons," 1906.
Courtesy Deutsches Museum, Munich.

Simple Models of/in a Complex Reality

This particular version of a reordering of the past from a present-centered viewpoint reflected processes happening in the wider public as well as the overall conceptual framework of the museum (which—then as today—is shaped by the power of the various actors and stakeholders involved). This can also be seen by the interpretation of the famous Michelson-Morley experiments.[17] Sticking to the prevalent accounts of the emergence of the theory of relativity Fuchs referred extensively to these famous experiments when he had to explain Einstein's theory in 1922 to the entire staff of the museum. As no original apparatus were available, the Deutsches Museum made a small model of the experiment in which the path of light was shown by strings (see figure 5.5). In the accompanying text the negative results of the "Michelson experiment" were declared to be a cornerstone of Einstein's theory of relativity.[18]

Whereas the historical linkages between relativity and the researches of Kaufmann and Bucherer were severed, a linkage between relativity and the Michelson-Morley experiments

FIGURE 5.5
Model of Michelson's experiment made in 1881. Courtesy Deutsches Museum, Munich.

was firmly established despite the limited historical basis for doing so.[19] The Michelson-Morley experiments were tightly tied to relativity and thus had a very precise meaning. But, this linkage could only work if the original experiments were "repositioned." First, the historical background of Michelson and Morley had to disappear. Second, one had to work out what these experiments "really meant" as seen from within the framework of relativity. And third, in order to be useful for educational purposes and to serve as a foundation of relativity theory, the long-lasting research project was reduced to a singular experiment (the one of 1881); all experimental difficulties and ambiguities were ignored and the outcome of the original experiment was presented as clear-cut and decisive. Like the model in the Deutsches Museum implicated in an ideal way: everything was fairly simple and straightforward. Take a light source, split the beam of light into two, which then travel at right angles to each other, each being reflected back upon itself by mirrors at the end of the path, and after crossing back and forth the two beams of light are reunited. The superposition will produce a specific pattern of fringes: If the velocity of light depends on its direction, one should observe a shift of interference fringes by turning the apparatus slowly, but Michelson and Morley could not detect any effect. This null-result could not be explained with the classical theory of an all-pervading, stationary electromagnetic ether, through which the earth moves.

Following traditional accounts, it was Einstein who solved this mysterious problem by introducing the special theory of relativity in 1905. The ether-drift experiment was turned into an "experimentum crucis" of relativity and at the same time into a means to explain the basic principles of Einstein's theory. All difficulties and potential problems of the experiments, all alternative explanations disappeared in order to provide a solid foundation of relativity.

However, in real life these experimental ambiguities had not disappeared. Things just were not as easy as the many oversimplified accounts of the Michelson-Morley experiments we still find today in textbooks and museums suggest. At least some physicists were still troubled by these experiments in the 1920s (i.e., nearly forty years after they had first been done!). One who was really angered by all these oversimplifications was the American physicist, Dayton Miller because he knew the Michelson-Morley experiments and its experimental problems only too well. According to his experience, the experiments were "the most trying and fatiguing, as regards physical, mental and nervous strain, of any scientific work." No wonder! He made in his experiments more than 100,000 readings, which involved walking in the dark for (in total) 100 miles in small circles around his apparatus to observe the interference fringes.[20]

In his presidential address to the American Physical Society on December 29, 1925 (which was later published in *Science*), Miller attacked the fixed meaning the Michelson-Morley experiments had in the mid-1920s.[21] For him it was only an "assumption that the ether-drift experiments of Michelson, Morley and Miller had given a definite and exact null result." This interpretation "had never been acceptable to him." He complained that the experiment "has always been applied to test a specific hypothesis," whereas he preferred to do extended observations "independent of any 'expected' result." He wanted to see the experiment not as an "experimentum crucis" but a precision measurement. In order to justify his approach he highlighted the fact that there had never been a "true null result" and emphasized the enormous difficulties of the experiments. By his observations at Mount Wilson he claimed to have found a "systematic displacement of the interference fringes of the interferometer corresponding to a constant relative motion of the earth and the ether at this observatory of ten kilometers per second."[22]

However, since the classical experiments by Michelson and Morley had achieved such a high status within the debate about relativity, Miller's results caused great excitement not only among physicists, but also in the wider public. Headlines in newspapers and popular science journals, sometimes with a question mark at the end, sometime without, announced: "Theory of relativity overthrown?"[23] We can get an impression of the reaction within the scientific community by looking, for example, at a letter of the physicist Jonathan Zenneck to his colleague Georg Joos. For Zenneck, the experimental results of Miller really were a "bad surprise," but he didn't dare to openly distrust them.[24] However, such debates were clearly beyond the grasp of the museum: With the simple string-model of the original Michelson-Morley experiment and the simplified account given in the accompanying text any visitor would have been at a loss to understand why on earth a renowned physicist like Miller had taken up again these questions of the past that had supposedly been solved long ago. For the museum, within which a very simple idea of how science works—"Question/theory–decisive answer/experiment–rejection of old theory and formulation of a new theory"—was advanced, reality proved to be too complicated and, of course, not only for the museum as its views reflected those of many scientists, historians, and philosophers of science at the time.

Arguments and Instruments

However, the Deutsches Museum got a second chance to do better, which I will now describe. The story starts with a particular reaction to Dayton Miller's experiments: a "repetition" by the German physicist Georg Joos.[25]

Joos wanted to show that there had been some error in Miller's experiment, but he, too, could not dismiss outright the detailed arguments about the problems of the measurement put forward by Miller. As a consequence he had to pay great attention to the issue of precision. As we know from other studies in our field, "precision" is a difficult notion and it is something the experimenter has to establish time and time again by careful argumentation.[26] Miller's detailed description of his painstaking experimental procedure is a good example; in a similar vein Joos constructed his argumentation in his major publication.[27] He first emphasized the overall sensitivity of the instrument that had been achieved by a light path of twenty-one meters. In order to make this sensitivity and precision visible at first sight for the wider public a photograph of the instrument accompanied many accounts of Joos's experiments, often a picture with a person nearby was chosen, thus enabling the reader to get a feeling for the size of the apparatus with a diameter of more than four meters (see figure 5.6).

However, although Joos's apparatus was as big as Miller's—and *looked* even bigger—what really matters is the total light path. And due to a smaller number of reflecting

FIGURE 5.6
The Joos interferometer in a basement room of the Zeiss company at Jena, ca. 1930. Courtesy Deutsches Museum, Munich.

mirrors Joos's light path was actually smaller than Miller's. Thus, Joos had to put forward more arguments in order to be convincing. So, he turned to a second major experimental problem: the temperature-sensitivity of the construction. Joos described in detail how he got in contact with the optical glass company, Schott, and how they tried to find an appropriate material on which to mount all optical components of the instrument. They agreed to use quartz, prepared in a new way. In the next step, Joos moved on to a major change he had introduced into the construction: Instead of letting the apparatus float on mercury (as was common in most experiments up to that time) Joos used a large number of springs to reduce the disturbing effects of vibrations. By means of these springs the four quartz plates were fixed onto the frame of the apparatus (see figure 5.7). The whole apparatus was carefully mounted onto the central axis and could be turned slowly (one turn in ten minutes) by an external electric motor. There was, however, one difficulty. Despite other plans made earlier during the planning of the experiment, Joos did his experiments in the end within a basement room of the Zeiss Company, although Miller had greatly emphasized the importance of working in an insubstantial building high on a hill. In order to justify this important divergence, Joos referred to the "core of the old ether theory" according to which any possible partial ether-drag would depend only on the mass of bodies. Since, as Miller had admitted, 1,800 meters height (i.e., 1,800 meters of atmospheric pres-

FIGURE 5.7
Detail of the Joos interferometer showing four mirrors (one adjustable), mounted on quartz-plates, which are attached to the frame by springs. Courtesy Deutsches Museum, Munich.

sure) would make no difference to the measurement, surely a little stone wall would not do so either.

So far, Joos had tried to counter each argument Miller had put forward; but this was not sufficient to convince others that he had done *better* than Miller. Therefore, Joos presented a new argument: He had substituted the direct observation of the interference fringes by an automatic camera. Thus, as Joos wrote, "documents" were produced which "could be verified by anyone."[28] When Zeiss handed the instrument over to the Deutsches Museum in 1935 it equated this photographic registration with an "absolute objective" measurement, and later in the museum the former direct observations were called "subjective"; of course, an argument often heard in the early twentieth century.[29] By such a detailed, technical description of the instrument in one of the most prestigious German journals, Joos hoped to convince everyone of the high precision he had achieved. To make an overall assessment easier he offered a comparison at the end of his paper: The precision achieved was as high as a measurement of the distance from Earth to the moon to a degree of accuracy of one centimeter. This comparison was then repeated in many accounts of Joos's experiments, for example in the journal of the Zeiss Company or later in the Deutsches Museum.[30] The well-known physicist Arnold Sommerfeld, who was informed about Joos's project, expected already in 1929 (i.e., before the experiments were actually done) an "experimental record of precision."[31] Joos expressed his hope that the precision reached would bring an end to all debates about any ether-drift.

However, parallel to this discussion about the precision of the experiment Joos was aware from the outset of the fact that he had to take the political aspects of his scientific enterprise into account. Writing to his friend and colleague Arnold Sommerfeld in December 1925 about the first planning stages, Joos remarked: "We want to handle the affair very discreetly, because otherwise obstructionism might occur. Among non-physicists here relativity theory is considered to be a subversive activity."[32] To Jonathan Zenneck he complained about the atmosphere at Jena and the strong influence of a "group of extreme nationalistic fools [völkische Narren]."[33] Given these political dimensions of the experiment, it comes as no surprise—in the context of the rise of National Socialism—that the strong linkage between the Michelson-Morley experiments and relativity theory now had to be severed. Hence, in Joos's publication in *Annalen der Physik*[34] in 1930 one looks in vain for the name "Einstein" or the term "relativity theory." Joos chose his words carefully and discussed his own work only in relation to the experiments of Michelson and Miller. He summarized his results at the end of the paper by simply stating: "Any potential ether-drift has to be smaller than 1.5 km/sec."[35] To some extent, the experiments were presented in a way very similar to Miller's five years earlier: They were reduced to a precision measurement but, most important, its wider implications remained hidden. Whereas Miller did so in order to find a niche for his research within the scientific community and within the tightly knitted net of arguments for the justification of relativity, Joos was concerned to present his research within a highly politicized atmosphere.

Only, in the liberal journal *Naturwissenschaften* was Joos more outspoken.[36] Here, he made clear that even a small ether-drift would bring an end to relativity theory and that he had proven that no such effect existed. In the various popular accounts of Joos's experiment we also find both versions, which shows how delicate this relationship between the experiment and Einstein's theory was in the early 1930s. An article in the house journal of Zeiss[37] portrayed the experiment as a precision measurement of the ether-drift without discussing its consequences but other popular science journals were more frank. The *Umschau*, for example, opens its article on Joos's experiments with the words: "In Einstein's special theory of relativity . . . " and thus makes explicit again the close connection between ether-drift experiments and Einstein's theory.[38]

Finding Room for the Joos-Interferometer in the Museum

In 1935, the interferometer was offered to the Deutsches Museum by the Zeiss Company.[39] Thus, the museum was now confronted with a set of questions: Where should it put the apparatus? How should it be related to the already existing exhibition? What should be written about it on the caption? Or more generally: What should it represent? The answers given to these questions reflected a mixture of very pragmatic concerns, the conceptual framework of the museum and, of course, the political dimensions of the apparatus as well. The museum could—and actually did—draw on the positioning already done by Joos and the Zeiss Company. Thus, the instrument traveled from Jena to Munich with the meaning already established at Jena in its baggage and both were "re-erected" together in the museum's gallery.

Since no other space was available for the large instrument, it was installed in the section on heat and energy. Parallel to the exchanges and negotiations with Zeiss the museum also got in contact directly with Georg Joos and requested a short description of his work. Joos sent a full page in small writing. Again, the experiment was explained only in relation to Michelson and the question of the existence of an ether-drift on an abstract level, technical details were emphasized, and any references to relativity theory and Einstein were omitted. Joos gave his result only in negative terms: "There is no ether."[40] The positioning by Zeiss and Joos of the instrument and the experiment provided the basis for the presentation in the museum, in the gallery as in guidebooks. The experiments were explained only in relation to the original Michelson-Morley experiments, the great precision of the instrument was highlighted, and the result of the experiment only given in the form: The movement of Earth has no effect on the velocity of light.[41] No further explanation of the meaning of such a statement was offered.

Although, according to an internal note, the museum considered Joos's description to be quite useless for practical purposes, it officially accepted it. In a letter to Joos, the museum only suggested the addition of one sentence to the effect that the instrument was an "exemplar of the highest mechanical and optical precision ever achieved in any physical instrument."[42] This addition was connected to a shift in the representation of the instrument in the museum. A similar phrase had already appeared in a letter from the museum to Zeiss in March 1935. Here, the museum defined the interferometer as a "hervorragendes Meisterwerke neuzeitlicher Präzisionsmechanik und -optik" ("outstanding masterpiece of modern precision mechanics and optics").[43] By using this terminology the museum was of course referring back to its own inception as a "Museum for the Masterworks of Science and Technology." Second, the apparatus could now be presented as an exemplar of a strong specific *German* tradition: precision mechanics and optics. The politically potentially problematic scientific dimensions of the experiment—the association with Einstein's theory of relativity—were thus moved into the background. Furthermore, the connection with the donor and maker of the instrument, the Zeiss Company, was strengthened, whereas the experiment and the experimenter—who might be considered to be more a political liability—became less important. In the text attached to the instrument Zeiss was consequently put first: It was a "Zeiss-Joos Interferometer" and in the first line of the text it was stated: "Zeiss built the apparatus in collaboration with Joos."[44]

Only years after the end of World War II, the linkage between the experiments and their relevance for the theory of relativity was reinstated in the museum.[45] The Joos interferometer remained in the provisional galleries which were opened in 1948 (it now stood in the astronomy section, see figure 5.8). For some years, it could still fulfill an important function that had been attached to it during its early career in the museum: the interferometer should serve as a signal to all visitors and interested parties that the museum did not only show antiquated objects but also kept up with the development of *modern* experimental physics. According to the curator, Franz Fuchs, the interferometer belonged to the group of "modern

FIGURE 5.8
The Joos interferometer in the provisional astronomy section at Deutsches Museum, 1953. Courtesy Deutsches Museum, Munich.

apparatus" that stood opposite "historical instruments."[46] For this reason the interferometer (supposedly) became "a main point of attraction in the physics gallery."[47] However, for how long could it continue to fulfill this function? To cut a long story short: Joos's apparatus was finally dismantled in 1957. Not only had the apparatus aged and the excitement about relativity abated, but, even more important, the concept for the new physics galleries had changed. It resembled much more what we would consider today a "'science-center" approach. Historical objects were seen as "dead objects." Complex measurement devices with a complicated history became an endangered species in the museum.[48]

Conclusion

This short sketch of the histories of the Joos interferometer and of the instruments by Kaufmann and Bucherer might be taken as a beginning of a detailed "biography of an instrument," or, as Roger Silverstone has put it, an analysis of how an object "gains its meaning through the various social, economic, political, and cultural environments through which it passes," and how its "passage can in turn illuminate those environments in the way that a flare or a tracer can illuminate the night sky."[49] Seen from this perspective, it is evident that museums are very important sites where new meanings of objects are constructed. Still, it is important to note that even within the restricted world of the museum, these meanings can change remarkably with time. The entry into a museum is not the end of the life of an object, it has an interesting life in the museum as well! Furthermore, the museum is not an isolated place outside of the world, but lives very much in it. Thus, the life of an object, the construction of its meaning within a museum clearly reflects such external influences, as, for example, the handling of Joos interferometer within the political situation in Germany demonstrates. These influences might come more directly or they might be mediated by policies and conceptual frameworks more specific to the institution itself. Still, what is striking (at least to me) is the *lack* of specific views on material objects developed within the museum. Despite being a museum with a richness of objects, no single standpoint with regard to the historical development of science, for example the emergence and rise of relativity, was formulated within the Deutsches Museum.[50] At the same time, the two main ideas in the Deutsches Museum about the history of science and its purposes—the concepts of "masterpieces" and of "genealogies of instruments"—additionally restricted the interpretation of artifacts.

To many of us this might not come as a surprise. To rephrase Jim Bennett, such displays of the material remains of past science may seem to us like bad history, but it was never intended to be good.[51] Science education (in a very specific form) has always exerted a strong influence on museums' policies and concepts, and it has been increasingly seen as being in conflict with a historical approach. This is true at least for the development of the physics galleries in the Deutsches Museum from the late 1950s onward when all the instruments discussed here have slowly disappeared from the galleries. Additionally, we have, of course, to take into account the development within the history of science in recent decades. Still, the question is whether museums absorbed many of these changes and developed a new approach. However, at least the three instruments mentioned here have been kept, and they could therefore be shown in 2005 in two different exhibitions on Einstein at Berlin and at Munich.[52] Now, they are back in the storerooms. When and how will they reappear? They wait for a better time when we will have hopefully learned more about analyzing and presenting their interesting lives in an adequate manner.

Notes

The title chosen for this written version of my presentation at the Artefacts conference in Utrecht in October 2004 refers, of course, to the famous novel by Charles Dickens, first published in 1854. Dickens' wonderful description of the "fact-oriented education" in an industrialized world always reminds me of the struggle with the so-called science education some people want us to do in museums. ("Now, what I want is, Facts. Teach these boys and girls nothing but Facts. Facts alone are wanted in life. Plant nothing else, and root out everything else. You can only form the minds of reasoning animals upon Facts: nothing else will ever be of any service to them."—that is the very beginning of the novel.)

1. The original German title is "Vorschläge zur Darstellung der Relativitätstheorie im Deutschen Museum." A typewritten script of the talk is kept in the Library of the Deutsches Museum: Fuchs, "Lichtbild-Vorträge, 1918–1936." Franz Fuchs was curator of physics in the Deutsches Museum from 1906 to 1951.

2. A short description of the events described in the following was given by Fuchs in "Der Aufbau der Physik im Deutschen Museum. 1905–1933." Fuchs included some quotations from his correspondence which has been (in part) preserved in the Archives of the Deutsches Museum (DMA) within the "Verwaltungsarchiv" (VA), mostly within the files associated with the physics department (VA 1838–1849). In the following I will give a summary of the events based on these documents. All translations of archival material are mine, unless otherwise stated.

3. Thirring, *Die Idee der Relativitätstheorie*. As he explained in the introduction of his book Thirring wanted to explain the core ideas of relativity without using any mathematics ("den gedanklichen Kern der Relativitätstheorie herauszuschälen und so gründlich darzustellen, als es bei völliger Vermeidung aller mathematischen Hilfsmittel möglich ist").

4. Thirring to Deutsches Museum. Thirring wrote: "Ich verhehle mir aber durchaus nicht, daß diese Tabelle, aus dem Zusammenhang meines Buches gerissen, dem Laien vollständig spanisch vorkommen muß. Wollte man nun eine erläuternde Legende dazu schreiben, so müßte man, um alles zu erklären, wiederum fast den ganzen Text des Buches bringen."

5. For a discussion of this development see, for example, Eckert, "Die Anfänge der Sommerfeld-schule"; Eckert, *Die Atomphysiker*; Jungnickel and McCormmach, *Intellectual Mastery of Nature*; Seth, *Principles and Problems*.

6. See, for example, Durant, *Museums and the Public Understanding of Science*; Butler, *Science and Technology Museums*; Hochreiter, *Vom Musentempel zum Lernort*; Pearce, *Exploring Science in Museums*, especially Arnold, "Presenting Science as Product or as Process"; Bennett, "Museums and the History of Science." See also the literature cited in ref. 9.

7. This has also important consequences for what is preserved and collected in museums as I have discussed in "Things That Once Were New Are Getting Old."

8. There already exists a body of interesting literature on the topic of the "biography of an object," see, for example, Alberti, "Objects and the Museum"; Hörning, "Vom Umgang mit den Dingen. Eine techniksoziologische Zuspitzung"; Kopytoff, "The Cultural Biography of Things"; Pearce, *Museums, Objects, and Collections*; Sichau, "Die Joule-Thomson-Experimente"; Silverstone, "The Medium is the Museum"; Kingery, *Learning from Things*; and the contributions in *History from Things*, edited by Lubar and Kingery.

9. Thus, if an instrument had been changed during its later lifetime, its usefulness for the exhibition was put in question. For example, an apparatus of the physicist E. Dorn offered to the Museum came to be regarded as almost useless for the Museum's purposes since Dorn had altered it in order to do new experiments. As his colleague Gustav Mie wrote in a letter to the Museum on November 30, 1921: "Diese neue Apparatur hängt also mit den Dingen, die Sie für das Deutsche Museum interessieren, eigentlich nicht zusammen."

10. The letter to Kaufmann signed by Fuchs dates from October 15, 1921; the answer by Kaufmann is from the October 27, 1921; here Kaufmann explained the situation to the Museum and offered the apparatus from 1906 (DMA, VA 1839). According to a Museum guidebook the apparatus (as the one by Bucherer, see below) was indeed on display in the 1920s; see Deutsches Museum, *Amtlicher Führer*. Both instruments reappeared in the 1960s in the new gallery on "Atomic Physics" where they were presented as predecessors of research in nuclear physics.

11. The apparatus came to the Museum in January 1922.

12. Kaufmann, "Über die Konstitution des Elektrons."

13. For a detailed discussion of these experiments see Miller, *Albert Einstein's Special Theory of Relativity*; Hon, "Is the Identification of Experimental Error Contextually Dependent?"

14. Einstein, "Über das Relativitätsprinzip."

15. Bucherer, "Die experimentelle Bestätigung des Relativitätsprinzips."

16. This is the very reason why the attention of historians of science has been drawn to them (see ref. 14).

17. For a very detailed account see Swenson, *The Ethereal Aether*.

18. Although the model made and presented in the Museum referred to the experiments done by Michelson and Morley in 1887 ("On the Relative Motion of the Earth and the Luminiferous Ether"), the Museum spoke only of Michelson in the accompanying text as well as in the catalog. See, for example, Deutsches Museum, "Amtlicher Führer." Further, in a text board in the Museum the decisive experiments are dated back to the year 1881—the time of the initial trials by Michelson—despite the fact that the model referred to the experiments in 1887 (see below).

19. Einstein started to link his theory of relativity with the Michelson-Morley experiments already in 1907 in his publication "Über das Relativitätsprinzip." For a discussion of these experiments and their later use in the justification of relativity see, for example, Buchwald, "The Michelson Experiment"; Collins and Pinch, *The Golem*; Hentschel, "Einstein's Attitude"; Kaiser, "Das Problem der entscheidenden Experimente"; and Stachel, *Einstein*.

20. Dayton C. Miller gave an account of his research in his publication "Significance of the Ether-Drift Experiments." For further information see the literature cited in refs. 18 and 20.

21. Miller, "Significance of the Ether-Drift Experiments." All following quotations are from this publication.

22. This velocity was much smaller than what originally had been expected. However, that is precisely why Miller wanted his experiments to be seen as a precision measurement and not as a test of a particular hypothesis or theory!

23. Reports were published, for example, in the popular science journals "Die Umschau" (Reichenbach, "Ist die Relativitatstheorie widerlegt?") and "Stein der Weisen" (Hoelling, "Erschütterung der Relativitätstheorie"). Newspapers like the "Berliner Tageblatt" (Kirchberger, "Die Grundlagen der Relativitätstheorie erschüttert?") and the "Jüdisches Familienblatt" (S. Jacoby, "Erneuter Angriff gegen Einsteins Relativitäts-Theorie") also published reports on Miller's claims, Einstein wrote a reply in the "Vossische Zeitung" ("Meine Theorie und Millers Versuche"). For further details see the article by Hentschel, "Einstein's Attitude towards Experiments."

24. Letter from Zenneck to Joos. The original German text is "Die Wiederholung des Michelson-Versuches ist allerdings eine böse Überraschung. Wenn sich alles bestätigt, woran ja eigentlich kaum zu zweifeln ist, so kann man wenigstens sagen, dass die ganze Aetherfrage in ein völlig neues Stadium gekommen ist." For further details about Georg Joos see below.

25. The career of Georg Joos is too complicated to give an adequate account here. In the following only some main aspects are stated. Joos was born in 1894 in south Germany. He studied physics first at the Technische Hochschule at Munich, did active service in the First World War, before taking up his PhD studies at the University of Tübingen. From 1921 until 1924 Joos was Professor Jonathan Zenneck's assistant at the physical institute of the Technische Hochschule at Munich. He left for Jena where he held a position as "außerordentlicher Professor" for four years before becoming full professor there. During his time at Jena he established strong links with the Zeiss Company. These contacts

enabled him to become head of research at Zeiss during the Second World War. After the war he became professor at the Technische Hochschule at Munich and member of the governing board of the Deutsches Museum. He died in 1959. Within Germany Joos was quite well-known for his text-book on theoretical physics ("Lehrbuch der Theoretischen Physik"); the first edition was published in 1932; nine editions followed until 1956.

26. For a discussion about "precision" and its (changing) meaning see for example, Wise, *The Values of Precision*; Cahan, *Meister der Messung*; Pyenson and Sheets-Pyenson, *Servants of Nature*.

27. There appeared several accounts of his experiments in various scientific and popular journals. The major publication was Joos, "Die Jenaer Wiederholung des Michelsonversuchs."

28. Joos, "Die Jenaer Wiederholung," 385. Joos paid much attention to the use of photography in science. In 1952, after the death of Ernst von Angerer, he even took in his hands the publication of the fifth and following editions of Angerer's book on scientific photography, *Wissenschaftliche Photographie. Eine Einführung in Theorie und Praxis*.

29. Carl Zeiss, Jena (company) to the Deutsches Museum, March 7, 1935. The term "subjective" was used in the text within the exhibition. A photograph of this text is preserved (DMA, BN 32258).

30. Anonymous, "Aetherwind?" 15. For the text used in the museum see ref. 30.

31. Sommerfeld to the University of Tübingen, June 11, 1929.

32. Joos to Sommerfeld, December 30, 1925. The full passage in the German original text is: "Vielleicht dürfte es Sie interessieren, daß wir hier mit Zürich zusammen ein Konsortium zur endgültigen Klärung des Michelson-Versuchs gegründet haben. Zeiß baut die Apparatur, die Schweiz übernimmt die Herrichtung eines geeigneten Raumes auf dem Jungfraujoch und kommt für die Transportkosten auf. Ich hoffe, die Apparatur in mancher Hinsicht zu verbessern, so will ich alles photographisch messen und mikrophotographisch auswerten. Die Sache ist noch einigermaßen im status nascendi aber die Hauptsache, die finanziellen Grundlagen, ist mit einem [annehmbaren?] Reservefonds geschaffen. Wir wollen aber hier die Sache ziemlich diskret behandeln, da sonst Quertreibereien möglich sind. Unter den Nichtphysikern hier gilt nämlich die Relativitätstheorie zum Teil als staatsgefährdend. Von der Schweiz werden voraussichtlich 2 Assistenten von E[dgar]. Meyer mitmachen. Hier werde ich mit Hilfe von Zeiß die Apparatur vorbereiten, die Versuche sollen dann in Jena, Zürich, Zuoz (Engadin) und Jungfraujoch ausgeführt werden."

33. Joos to Zenneck, November 27, 1925.

34. Joos, "Die Jenaer Wiederholung."

35. Joos, "Die Jenaer Wiederholung," 407.

36. Joos, "Wiederholungen des Michelson-Versuchs." Joos made a direct reference to Einstein and relativity theory also in "Die Jenaer Nachprüfung des Michelsonschen Ätherwindversuchs."

37. Anonymous, "Aetherwind?"

38. Saller, "Gibt es einen Aetherwind?—Nein"

39. Zeiss wrote a letter to the Museum on March 7, 1935, offering the instrument. This seems to be the starting point of the negotiations. Zeiss explained that too many visitors wanted to see the apparatus and they, as a company, could not handle this interest. Already on March 30, the packed instrument arrived in Munich where two engineers from Zeiss helped to re-erect it in the physics gallery within two weeks.

40. Joos to the Deutsches Museum, May 31, 1935. Here, Joos followed the version in the third edition of his textbook on theoretical physics (see ref. 26): "The experiment has decided against the existence of the ether or absolute space."

41. See, for example, Conzelmann, *Rundgang durch die Sammlungen*, 108.

42. Deutsches Museum to Joos, June 5, 1935.

43. Deutsches Museum to Carl Zeiss, Jena (company), Jena, March 14, 1935.

44. Copy of the text used in the exhibition in the Deutsches Museum (see ref. 30). The phrase used in the museum echoes, of course, the opinion expressed by Arnold Sommerfeld in 1929—but with this slight shift discussed here: Sommerfeld's praise was directed toward Joos as an experimenter, not to the manufacturer of the instrument, Zeiss. Whether Sommerfeld, who acted as an adviser to the Deutsches Museum, played a role in the erection of the apparatus in 1935, I do not know at present.

45. There is a typewritten script—probably intended for publication—by Franz Fuchs, dated September 8, 1948, in which the Joos interferometer was mentioned. In this context it was remarked that the Michelson experiments were a basis for the emergence of Einstein's theory of relativity ("Es war dies eine der Grundlagen für die Aufstellung der Einsteinschen Relativitätstheorie"). In the guidebook published in 1956 the Joos interferometer was also referred to (Deutsches Museum, *Kurzer Rundgang durch die Sammlungen*, 27). It was declared that Joos proved with this apparatus with great precision that the velocity of light did not depend on the velocity of the light source. To this the half-sentence was added to the effect: "what is relevant for the theory of relativity" ("was für die Relativitätstheorie von Bedeutung ist").

46. Fuchs, "Die naturwissenschaftlichen Gruppen," 468. To keep up with the development of science and technology had been an important issue for the Deutsches Museum. In 1938, for example, Jonathan Zenneck, head of the Deutsches Museum from 1933 to 1953, visited the Conservatoire des Arts et Métiers in Paris. He was not impressed, to the contrary. According to his opinion the Conservatoire had become like a "dead tree" because it had stopped growing. It had not taken in the latest developments of science. Thus, for Zenneck it was nothing but a collection of antiquared objects. In order to avoid such a development the Deutsches Museum should constantly try to bring in new objects within its galleries (Zenneck, "Die Anpassung"). For more details: Sichau, "Things That Once Were New Are Getting Old."

47. Meißner, "Zum 60. Geburtstag von Georg Joos," 193.

48. Although we can discern quite a shift of the conceptual framework, to some extant there is also a persistence of what I have loosely labelled the "traditional approach" within the Deutsches Museum. In the planning documents from the late 1950s and 1960s on the new physics galleries—which are the source of the quotations given in the text—it is stated repeatedly that a historical approach would only complicate matters, that real instruments would be too difficult to be comprehended and would not help to understand the laws of nature and so on; for further details see Sichau, "Things That Once Were New" and "Meisterwerke und Entwicklungsreihen." These sentiments expressed by former curators at the Deutsches Museum reflect the ambiguous position of the physics galleries, the struggle between a historical approach and a "science-center" approach. With regard to the rise of the latter in recent years, Jim Bennett has pointed out clearly its implications: "Where the new ambitions demanded clear and de-contextualized presentation of 'the science,' objects seemed ambiguous and contingent. Insufficiently malleable to the new mission, they retained too much of their own agenda. . . . Galleries full of them seemed to raise too many questions where what was wanted were answers" (Bennett, "Museums and the History of Science," 606).

49. Silverstone, "The Medium is the Museum."

50. It is especially remarkable that museums have not been more proactive in discussing the relationship between science and technology although, given their collections, they would have an ideal starting point for such a discussion.

51. Bennett, "Museums and the History of Science," 606.

52. Catalogs of these exhibitions have been published: (1) For Munich: Brachner, Hartl, and Sichau, *Abenteuer der Erkenntnis* (chapter 7 is on the Joos-interferometer); (2) For Berlin: Renn, *Albert Einstein—Ingenieur des Universums. Einsteins Leben und Werk im Kontext* (a collection of essays in connection with this exhibition has been published separately as Renn, *Albert Einstein—Ingenieur des Universums. Hundert Autoren für Einstein*).

References

Alberti, Samuel. "Objects and the Museum." *Isis* 96 (2000): 559–71.

Anonymous. "Aetherwind?" *Zeiss-Werkzeitung* 1 (1932): 14–15.

Arnold, Ken. "Presenting Science as Product or as Process: Museums and the Making of Science." In *Exploring Science in Museums*, edited by Susan Pearce. London: Athlone, 1996.

Bennett, Jim. "Museums and the History of Science—Practitioner's Postscript." *Isis* 96 (2005): 602–8.

Brachner, Alto, Gerhard Hartl, and Christian Sichau, eds. *Abenteuer der Erkenntnis—Albert Einstein und die Physik des 20. Jahrhunderts*. Munich: Deutsches Museum, 2005.

Bucherer, Alfred H. "Die experimentelle Bestätigung des Relativitätsprinzips." *Annalen der Physik* 28 (1909): 513–36.

Buchwald, Jed Z. "The Michelson Experiment in the Light of Electromagnetic Theory before 1900." In *The Michelson Era in American Science 1870–1930*, edited by S. Goldberg and R. Stuewer. New York: American Institute of Physics, 1988.

Butler, Stella. *Science and Technology Museums*. Leicester: Leicester University Press, 1992.

Cahan, David. *Meister der Messung. Die Physikalisch-Technische Reichsanstalt im Deutschen Kaiserreich*. Weinheim: VCH, 1992.

Carl Zeiss Company to the Deutsches Museum, March 7, 1935. Archives of the Deutsches Museum (hereafter DMA), VA 1878.

Collins, Harry, and Trevor Pinch. *The Golem—What Everyone Should Know about Science*. Cambridge: Cambridge University Press, 1993.

Conzelmann, Theodor, ed. *Rundgang durch die Sammlungen*, 4th ed. Munich: Deutsches Museum, 1938.

Deutsches Museum to Carl Zeiss, Jena (company), March 14, 1935. DMA, VA 1878.

Deutsches Museum to Georg Joos, June 5, 1935. DMA, VA 1877.

Deutsches Museum. *Amtlicher Führer durch die Sammlungen*, 2nd ed. Munich: Deutsches Museum, 1928.

Deutsches Museum. *Kurzer Rundgang durch die Sammlungen*. Munich: Deutsches Museum, 1956. DMA, VA 1901.

Dickens, Charles. *Hard Times for These Times*. London: Bradbury & Evans, 1854.

Durant, John, ed. *Museums and the Public Understanding of Science*. London: Science Museum, 1992.

Eckert, Michael. "Die Anfänge der Sommerfeldschule, 1907–1912." In *Arnold Sommerfeld. Wissenschaftlicher Briefwechsel*. Bd. 1: 1892–1918, edited by M. Eckert and K. Märker. Berlin: GNT, 2000.

Eckert, Michael. *Die Atomphysiker. Eine Geschichte der theoretischen Physik am Beispiel der Sommerfeldschule*. Braunschweig: Vieweg, 1993.

Einstein, Albert. "Meine Theorie und Millers Versuche." *Vossische Zeitung* (January 19, 1926).

Einstein, Albert. "Über das Relativitätsprinzip und die aus demselben gezogenen Folgerungen." *Jahrbuch der Radioaktivität und Elektronik* 4 (1907): 411–62.

Fuchs, Franz, to Walter Kaufmann, October 15, 1921. DMA, VA 1839.

Fuchs, Franz. "Die naturwissenschaftlichen Gruppen im Deutschen Museum." *Naturwissenschaftliche Rundschau* 10 (1950): 468–69.

Fuchs, Franz. "Vorschläge zur Darstellung der Relativitätstheorie im Deutschen Museum." In Library of the Deutsches Museum: Fuchs, "Lichtbild-Vorträge, 1918–1936."

Fuchs, Franz. Untitled script, September 8, 1948. DMA, VA 1901.

Hentschel, Klaus. "Einstein's Attitude towards Experiments: Testing Relativity Theory." *Studies of the History and Philosophy of Science* 23 (1992): 593–624.

Hochreiter, Walter. *Vom Musentempel zum Lernort.* Darmstadt: WBG, 1994.

Hoelling, J. H. "Erschütterung der Relativitätstheorie." *Stein der Weisen Heft* 1 (1927): 11–13.

Hon, Giora. "Is the Identification of Experimental Error Contextually Dependent? The Case of Kaufmann's Experiment and Its Varied Reception." In *Scientific Practice. Theories and Stories of Doing Physics*, edited by J. Buchwald. Chicago: University of Chicago Press, 1995.

Hörning, Karl. "Vom Umgang mit den Dingen. Eine techniksoziologische Zuspitzung." In *Technik als sozialer Prozeß*, edited by P. Weingart. Frankfurt: Suhrkamp, 1990.

Jacoby, Siegfried. "Erneuter Angriff gegen Einsteins Relativitäts-Theorie." *Jüdisches Familienblatt* 7 (1926).

Joos, Georg. "Die Jenaer Nachprüfung des Michelsonschen Ätherwindversuchs." *Unterrichtsblätter für Mathematik und Naturwissenschaften* 6 (1933): 185–89.

Joos, Georg, to Arnold Sommerfeld, December 30, 1925. DMA HS 1977-28/A: 157.

Joos, Georg, to Deutsches Museum, May 31, 1935. DMA, VA 1877.

Joos, Georg, to Jonathan Zenneck, November 27, 1925. DMA, NL 053.

Joos, Georg. "Die Jenaer Wiederholung des Michelsonversuchs." *Annalen der Physik* 7 (1930): 385–407.

Joos, Georg. "Wiederholungen des Michelson-Versuchs." *Die Naturwissenschaften* 38 (1931): 784–89.

Jungnickel, Christa, and Russell McCormmach. *Intellectual Mastery of Nature: Theoretical Physics from Ohm to Einstein.* Chicago: University of Chicago Press, 1986.

Kaiser, Walter. "Das Problem der entscheidenden Experimente." *Berichte zur Wissenschaftsgeschichte* 9 (1986): 109–25.

Kaufmann, Walter, to Deutsches Museum, October 27, 1921. DMA, VA 1839.

Kaufmann, Walter. "Über die Konstitution des Elektrons." *Annalen der Physik* 4 (1906): 487–553.

Kingery, W. D. *Learning from Things: Method and Theory of Material Culture Studies.* Washington, DC: Smithsonian Institution Press, 1996.

Kirchberger, P. "Die Grundlagen der Relativitätstheorie erschüttert?" *Berliner Tageblatt* 16 (January 10, 1926).

Kopytoff, Igor. "The Cultural Biography of Things." In *The Social Life of Things: Commodities in Cultural Perspective*, edited by Arjun Appadurai, 64–91. Cambridge: Cambridge University Press, 1986.

Lubar, S., and W. D. Kingery, eds. *History from Things: Essays on Material Culture.* Washington, DC: Smithsonian Institution Press, 1993.

Meißner, W. "Zum 60. Geburtstag von Georg Joos." *Zeitschrift für angewandte Physik* 6, (1954): 193–94.

Michelson, Albert A., and Edward W. Morley. "On the Relative Motion of the Earth and the Luminiferous Ether." *American Journal of Science* 34, no. 203 (1887): 333–45.

Mie, Gustav, to Deutsches Museum, November 30, 1921. DMA, VA 1839.

Miller, Dayton C. "Significance of the Ether-Drift Experiments of 1925 at Mount Wilson." *Science* 58 (1926): 433–43.

Miller, Arthur I. *Albert Einstein's Special Theory of Relativity. Emergence and Early Interpretations (1905—1911)*. Reading, MA: Addison-Wesley Pub. Co., Advanced Book Program, 1981.

Pearce, Susan, ed. *Exploring Science in Museums*. London: Athlone, 1996.

Pearce, Susan. *Museums, Objects, and Collections: A Cultural Study*. Leicester: Leicester University Press, 1992.

Pyenson, Lewis, and Susan Sheets-Pyenson. *Servants of Nature. A History of Scientific Institutions, Enterprises and Sensibilities*. London: Fontana, 1999.

Reichenbach, Hans. "Ist die Relativitätstheorie widerlegt?" *Die Umschau* (April 24, 1926): 325–28.

Renn, Jürgen, ed. *Albert Einstein—Ingenieur des Universums. Einsteins Leben und Werk im Kontext*. Weinheim: Wiley-VCH, 2005.

Renn, Jürgen, ed. *Albert Einstein—Ingenieur des Universums. Hundert Autoren für Einstein*. Weinheim: Wiley-VCH , 2005.

Saller. "Gibt es einen Aetherwind?—Nein." *Die Umschau* 45 (1931): 894–96.

Seth, Suman. *Principles and Problems: Constructions of Theoretical Physics in Germany, 1890–1918*. PhD Dissertation, Princeton University, 2003.

Sichau, Christian. "Meisterwerke und Entwicklungsreihen—Selbstvergewisserungen und Positionierungen im Deutschen Museum." Paper presented at the 87th Annual Conference of the DGGMNT, Mainz, September 24–27, 2004.

Sichau, Christian. "Things That Once Were New Are Getting Old—And Other Problems of 20th Century Science in Museums." Paper presented at the workshop "Curating 20th Century Science." Utrecht, October 17–18, 2005.

Sichau, Christian. "Die Joule-Thomson-Experimente: Anmerkungen zur Materialität von Experimenten." *Zeitschrift für Geschichte der Naturwissenschaften, Technik und Medizin* 8 (2000): 222–43.

Silverstone, Roger. "The Medium is the Museum: On Objects and Logics in Times and Spaces." In *Museums and the Public Understanding of Science*, edited by John Durant. London: Science Museum, 1992.

Sommerfeld, Arnold, to the University of Tübingen, June 11, 1929. DMA, NL 89.

Stachel, John. *Einstein from 'B' to 'Z'*. Basel: Birkhäuser, 2002.

Swenson, Loyd. Jr. *The Ethereal Aether: A History of the Michelson-Morley-Miller Aether-Drift Experiments, 1880–1930*. Austin: University of Texas Press, 1972.

Thirring, Hans, to the Deutsches Museum, March 24, 1924. DMA, VA 1846.

Thirring, Hans. *Die Idee der Relativitätstheorie*. Berlin: Springer, 1921.

von Angerer, Ernst. *Wissenschaftliche Photographie. Eine Einführung in Theorie und Praxis*, enlarged 5th edition by Georg Joos. Leipzig: Akademische Verlagsgesellschaft, 1952.

Wise, M. Norton, ed. *The Values of Precision*. Princeton, NJ: Princeton University Press, 1995.

Zenneck, Jonathan, to Georg Joos, November 24, 1925. DMA, NL 053.

Zenneck, Jonathan. "Die Anpassung von naturwissenschaftlichen und technischen Museen." *Technik-Geschichte* 30 (1941): 143–48.

CHAPTER 6

Why Display?

REPRESENTING HOLOGRAPHY IN MUSEUM COLLECTIONS

Sean F. Johnston

Introduction

Magazine journalists, museum curators, and historians sometimes face similar challenges in making topics or technologies relevant to wider audiences. To varying degrees, they must justify the significance of their subjects of study by identifying a newsworthy slant, a peda-gogical role, or an analytical purpose. This *chasse au trésor* may skew historical storytelling itself. In science and technology studies, the problem potentially is worst for those subjects that do not retain an enduring capacity for spectacle or do not readily conform to templates of progress or economic importance.

The case of holography is an excellent illustration of these points. An unusually wide-ranging subject, holography has attracted competing interpretations of intellectual novelty, technological application, and cultural significance. This chapter examines the history of museum collections and exhibits of holography from the explosion of interest in the technique in 1964 to the end of the twentieth century. I argue that holography presents a particular challenge to museum collections and to popular representations of the history of science and technology.

A Historical Sketch of Holography

Historical treatments are, of course, as susceptible as any other kind of summary to authorial bias. In the following, though, I will emphasise the manifold interpretations and contested success of the subject, themes developed at length elsewhere.[1]

Some of the intellectual roots of holography could be traced back to studies of the wave nature of light in the late nineteenth and early twentieth centuries, although this retrospective identification is contentious.[2] Practical applications of this conception were worked out over decades, for example, in Ernst Abbe's concept of spatial filtering in microscopy during the 1870s and the technique of phase contrast microscopy developed by Frits Zernike during the late 1930s.[3] Notions of two-step imaging—much later deemed to be an organizing principle of holography—were also circulating during the interwar period, for example, by Polish physicist Mieclslav Wolfke in 1920 and Sir Lawrence Bragg in 1939. The first investigations that stressed these ideas, however, were undertaken by Dennis Gabor, a Hungarian physicist and electrical engineer working at British Thomson-Houston in Rugby, England, in 1947.

Gabor conceived that an optical image could be reproduced using coherent light (that is, light of well-defined wavelength and phase) by a two-step process. First, an interference pattern would be recorded on photographic film by light diffracted around the microscopic object; second, this pattern would be used to diffract a subsequent beam of coherent light which, unintuitively, would focus to form a three-dimensional image of the original object. While elegant, Gabor's so-called wavefront reconstruction principle nevertheless was unimpressive to all but a handful of researchers in the UK, California, and Germany. His enunciation of the theory was arcane and alien even to mentors who were versed in the high science of optics such as Bragg and Max Born; the interference pattern, or *hologram*, generated an image that was polluted by a second, defocused image; it yielded unimpressive reconstructions poorer in quality than the direct electron microscope imagery that Gabor sought to improve; and, while in principle it was a three-dimensional technique, the available light sources—usually a filtered mercury lamp shining through a pinhole—were impracticable for the purpose. Within a decade all had given up on the idea.[4]

Elements of Gabor's idea were nevertheless independently reinvented by two other investigators. At the Vavilov State Optical Institute in Leningrad, the largest optical research center in the Soviet Union, Yuri Denisyuk, a young researcher pursuing a *Kandidat* degree, began to explore the possibilities of recording a wavefront of light. Between 1958 and 1961 he developed what he called *wave photography*, and a technique to record the wavefront of light through the depth of a photographic emulsion. The result was a reflecting plate that reproduced the optical effects of shallow mirrors or reflective rulers. Denisyuk struggled to express the theoretical underpinnings and wider significance of his wave photography and gained little attention at home or abroad when the work was published.[5]

At the University of Michigan's Willow Run Laboratories (WRL), dedicated to military contract research from the late 1940s, a major development program for synthetic aperture radar led to the third elaboration of Gabor's concept. Emmett Leith, a research engineer at WRL, conceived a relationship in 1956 between the analysis of synthetic aperture radar signals and the radar image derived from them. He merged communication theory, familiar to the electrical engineers at the lab, with the optics of diffraction and interference, so-called physical optics. This fruitful union permitted a successful extension of Gabor's ideas and, by 1963, Leith and colleague Juris Upatnieks had begun using newly available lasers as a source of coherent light. The result was stunningly accurate reproduction of three-dimensional objects, as well as a host of other capabilities having military significance for the optical processing of information. Despite the classified aims of WRL, Leith's and Upatnieks' technique was publicized by the American Institute of Physics as *lensless photography*, a label that the researchers were pleased to take up in a round of presentations at engineering conferences in America over the following two years.[6] Early international scientific contacts were courted, and a rival history promulgated, by another University of Michigan scientist (1963–1967), Professor George W. Stroke, who founded the Electro-Optical Sciences Laboratory in the Electrical Engineering Department, a competition that had a bearing on the subsequent award of the Nobel Prize to Gabor alone in 1971.[7]

Some two dozen researchers at WRL, supported by military contracts, began to pursue the implications of lensless photography, as did other researchers around the world.[8] Within a year, several groups (notably at the National Physical Laboratory in Teddington, UK) had invented variants of *holographic interferometry*, a technique of employing optical fringes to signal movement, deformation, or vibrational modes of solid objects.[9]

Following the publicity of the American work, Denisyuk's research was rehabilitated in the Soviet Union, thanks in no small part to the intervention of prominent physicists Petr Kapitsa and Mstislav Keldysh, president of the Soviet Academy of Sciences. By early 1966 the term *holography* encapsulated the newly conceived young science, and *holographer* became the label adopted by its specialists.

Large American firms either associated with military contracts or having commercial research labs, such as TRW, Sperry Rand, Lockheed, Hughes Aircraft, Rockwell, Bell Telephone, RCA, and CBS Labs, began to fund holography research during the late 1960s. Even earlier, however, some smaller firms had begun to explore the commercial possibilities for sales based on holograms or on consultation for holographic analysis techniques. Among the most important early contributors was the Conductron Corporation in Ann Arbor, Michigan, which was populated by a number of former Willow Run and University of Michigan technical workers. By 1967 the company had mass-produced a production run of a half-million holograms, developed a pulsed laser to record human portraits, and investigated new types of holograms that could be viewed in white (i.e., noncoherent) light. For investors and potential customers, the company and its founder, Keeve M. "Kip" Siegel, promoted holography as an expression of inexorable progress, in which three-dimensional television and home movies would be commonplace within a decade.[10] Its pursuit of markets, however, was largely unsuccessful: advertisers, Detroit car manufacturers, medical companies, and publishers seldom returned for repeat orders, and popular response was muted, for reasons the engineers struggled to understand.[11]

By the end of the decade, however, new users beyond military and academic research networks were beginning to take up holography. Artisans reconceived the hologram as medium for personal expression and as a product for a potential cottage industry. In the San Francisco Bay area, Lloyd Cross, a former Willow Run engineer, attracted a coterie of art-school graduates and others in a communal atmosphere to form the first School of Holography in 1971 and, shortly afterward, the Multiplex Company, which developed a new and soon widespread form of hologram.[12] Their "multiplex" or *holographic stereogram* combined movie footage with individual strip holograms arranged as a cylinder to create short holographic movies of subjects ranging from molecular models to erotica. They also incorporated a new hologram geometry developed by Polaroid researcher Stephen Benton, known colloquially as the *rainbow hologram*, to allow their multiplex hologram to be viewed with an unfrosted light bulb.

The San Francisco School, and others like it in America in the early 1970s and in Western Europe by the end of the decade, emphasised new understandings of the subject. Building on countercultural roots, holography was reinvented as an expression of holism, as a medium amenable to new forms of art and aesthetics, and as a concept and technology that could transform society. These wider interpretations were supported by the speculations of physicist David Bohm and psychophysiologist Karl Pribram.[13]

The expectations of scientific holographers, supported by corporate and military funding, and of artisans, pursuing cottage-industry ideals, were that holography would progress in an inevitable fashion, evolving toward higher quality, lower cost, and ubiquitous imaging—in effect, a combination of a positivist interpretation of the underlying scientific knowledge and a progressivist anticipation for the corresponding technology. But a third community, appearing at about the same time as artisanal holographers, stressed the aesthetic potential of holograms for fine art. The first artists to explore holography around 1968–1969 collaborated with university or commercial holographers, and included Bruce Nauman (working briefly with Conductron engineers), Margaret Benyon (increasingly self-reliant at a variety of British institutions), and Harriet Casdin-Silver in Boston, initially collaborating with scientists at American Optical, Polaroid, and Brown University. They were followed by a wave of holographic artists who shunned scientific connections, trained at schools of holography in San Francisco, Chicago, and New York and, from the 1980s, at European schools and short-lived institutional homes for aesthetic holography such as the Royal College of Art in London (1985–1994). Scientific, artisanal, and aesthetic holographers comprised three relatively immiscible and durable communities, although all three declined in numbers during the 1990s.[14]

Despite the failure of companies such as CBS and Conductron to identify viable market niches in the late 1960s, a cottage industry for holograms grew slowly during the following decade. These products were triggered and nurtured by large-scale public exhibitions of holograms. The first isolated holograms had been displayed in university departments or company foyers during the 1960s, with the first public shows held in a handful of American and British art galleries from 1968. Over the following decade, however, such exhibitions garnered increasingly large audiences around the world: Holography 1975: The First Decade (New York, 1975); Holografi: Det 3-Dimensionel Mediet (Stockholm, 1976); Whole Message (Vancouver, 1976); Light Fantastic (London, 1977 and 1978); Alice in the Light World (Tokyo, 1978); Holographie Dreidimensionale Bilder (Berlin, 1979); Olografia (Rome, 1979); Light Dimensions (Bath, 1983); Images in Time and Space (Montreal, 1986). Exhibition visitors rose from a few thousand to a few hundred thousand over a decade. Soviet holographers mounted their own displays in galleries, buses, and trains, estimating that a million visitors viewed Ukrainian and Russian exhibitions each year during the early 1980s, and also exported traveling exhibitions to venues in Europe and Asia.

Numerous small hologram manufacturing firms and sales outlets sprang up in the wake of such exhibitions, particularly in Western Europe, but none of them nurtured an enduring public interest in the medium. Commercial holograms seemed to follow a downward spiral in size, price, and aesthetic value; within a few years most hologram shops had discovered that viable sales could be sustained only by national distribution of low-cost products to science centers and the like.

An important factor in the decline of a commercial cottage industry was the development of the embossed hologram. In this variant, the fine fringe patterns are recorded as surface relief—variations in the thickness of the underlying medium on which reflective film is deposited, rather than as variations in transmission as in photographic emulsions. An embossed plastic tape scheme had been developed by RCA during the early 1970s for an aborted video player project dubbed Selectavision Holotape, and a photopolymer process was devised at the same time by Holotron, another American company that was to acquire most of the significant holography patents but proved unable to profit from them. The schemes were further developed from 1979, however, by three individuals: Mike Foster, a former light show operator, Steve McGrew of Light Impressions, and Ken Haines, a former WRL and Holotron engineer.

When embossed holograms were promoted for credit cards and magazine covers from the early 1980s by Ed Weitzen of American Bank Note Holographics (ABNH), the commercial market for conventional photographic-emulsion-based holograms began to decline perceptibly. Other materials for generating higher quality holograms, including photopolymers developed by DuPont and by Polaroid, and dichromated gelatine (DCG) developed by artisanal holographers to create bright artworks, and by scientific holographers principally for military applications such as Head Up Displays (HUDs), also failed to make substantial inroads owing to their intrinsically higher cost, lower volume, and relatively complex handling procedures compared to embossed holograms, which could be manufactured in volume on little-modified printing presses. Ironically, the optical quality of the increasingly ubiquitous embossed holograms was far inferior to their predecessors because of the flexibility of the backing material and uncontrolled lighting. As a result, manufacturers increasingly simplified the imagery of embossed holograms through the 1980s, relegating their use to repetitive patterns and shallow, uncritical images suitable for packaging or attention-getting displays.

Art holography had emerged slightly too late to ally itself with the art and technology movement of the 1960s, and it was also rebuffed from the mid-1970s by art critics who classed the medium as unimaginative and technologically oriented. Holographic art had its strongest institutional support in engineering centers, notably the Massachusetts Institute of Technology (MIT), and with patronage by a few institutions during the tenures of certain employees. The longest-lived support for art holography, however, came from the New York Museum of Holography (MoH, 1976–1992).

Holography in Museums

To a first approximation, permanent museum representations of holography evolved from the short-lived exhibitions of the 1970s. In most cases, the lineage was direct.

The New York Museum of Holography (MoH), the first and arguably most influential of these early venues, was a private initiative later supported by admission fees, New York State grants, and private donations. It grew alongside the rise in popular interest in holography, and in fact its originators, Joseph "Jody" Burns and Rosemary "Posy" Jackson, were responsible for the first large New York and Stockholm exhibitions. As the first enduring showcase for holograms, the MoH also purchased art holograms, thereby supporting the nascent community of artists, and instituted an artists in residence (AIR) program to provide access to the expensive technology and to encourage a sharing of skills. The first director of the MoH, Posy Jackson, left the organization in 1983, but she continued to support it financially through a family fund, the Shearwater Foundation. Shearwater was to be the principal institutional supporter of art holography over the following two decades until its closure in 2004.

The MoH sought to become a world center for the subject by seeking to collect a full and representative collection of holograms, by mounting both permanent and temporary exhibitions highlighting the artistic, scientific, and commercial applications of the medium, and by continuing to manage traveling exhibitions around America and Europe. The museum also attempted to build an international reputation by protecting its name from European competition—notably from the Museum für Holographie und neue visuelle Medien in Pulheim, Germany (founded by Matthias Lauk in 1979), the Musée de l'Holographie in Paris (founded 1980), and the Fine Arts Research and Holographic Center in Chicago (founded 1982)—all of which resisted, arguing that the words "museum" and "holography" were hardly unique. The MoH also exported the American model abroad, playing a significant role in defining the new holography program at Goldsmith's College, London, in 1980, and in complaints about being "appalled at the geographic focus" of a television broadcast in the Horizon/Nova series that had portrayed developments in British holography.[15]

Thus the MoH actively sought to define the public perception of holography. Indeed, public education proved to be an activity that sapped the resources of the MoH. While school tours boosted admissions income for the small museum (which had annual attendances ranging up to 60,000 in the late 1970s but falling to 26,000 near its demise in the early 1990s), museum personnel found themselves acting as spokespersons for the young holography industry, boosters for an uncertain art market and information sources for school essays around the country. None of these latter activities augmented museum income.[16]

The New York Museum of Holography and its variants in other countries sought financial viability and, not unsurprisingly, developed exhibitions to highlight publicly appealing themes that had brought crowds to temporary exhibitions. The MoH promoted traveling exhibitions, most of which emphasised a historic dimension or chronological development. Consequently, the staff requested "historic" holograms from their creators to represent the lineage of the medium, and the resulting collecting pattern was opportunistic.[17] In effect, the historiography popularized by the museum was based on the technological progressivism that dominated the holography industry and perceptions of public expectations concerning the medium. These themes were often dissociated from the views held by practicing scientists and, indeed, lacked historical evidence.

Holography in Other Museum Venues

The role adopted by the MoH and its European counterparts was later assumed by other institutions. Most of these were either temporary exhibits or limited in scale, typically at science centers or the equivalent. Several cases, however, are worthy of further discussion.

In common with the MoH, Matthias Lauk's Museum für Holographie produced traveling hologram shows, displayed in banks and town halls, cultural centers, trade fairs, and museums, especially in Germany. Like a number of other small holography businesses, the Museum für Holographie supplemented income from visitors and hologram sales with holography classes beginning in 1984. That year, Lauk's private collection of holograms, still the basis for the museum, was sold to the newly founded Zentrum für Kunst und Medientechnologie (Center for Art and Media Technology).[18] This German state institution was almost unique in conserving and collecting holograms, being paralleled on a much smaller scale by the Victoria & Albert Museum in London and contrasting with the National Museum for Photography, Film & Television (later the National Media Museum) in Bradford, UK.

Nevertheless, the private museums of holography struggled. The New York Museum of Holography was fighting bankruptcy by 1991 in the face of declining visitor numbers and other financial support. Others saw the challenges in deeper terms, in the very categorization of the subject. Glenn Wood, for instance, marketing manager for Ilford Ltd., responsible for holographic products since 1983, suggested that holography had to merge with up-and-coming technologies to retain popular interest in the early 1990s. He advised the final director of the New York Museum of Holography to

> drop holography as a name and to link the museum instead with virtual reality: If you could somehow hook onto it pointing out that the hard copy version has been around for some time and is called holography maybe you could get something going that way. There are also many similarities between the Los Angeles crowd involved in Virtual Reality (VR) and holographers (like they are all crazy people). . . . Change the MoH to the MOVR and relaunch. I don't think anyone gets excited about holography anymore, at least not the corporations or institutions you need to stay viable.[19]

Both the MoH and Ilford holography division closed in 1992 and those decisions were the first of a wave. The Museum für Holographie und Neue visuelle Medien, founded in late 1979, closed in 1994; the Holography Unit of the Royal College of Art, an important source of postgraduate fine-art holographers, closed the same year; the Canada Council ceased funding for art holography in 1995, and Light Impressions of California, the first firm to market embossed holograms, closed operations owing to inadequate sales; in 1996 Agfa Gavaert, long the dominant supplier of silver-halide holographic emulsions, announced that it would end production; the final Gordon Research Conference of scientist-holographers, an important American series of meetings initiated for holography in 1972, was held in 1997; and, also in 1997, Hughes Power Products, a division of Hughes Aircraft, ceased photopolymer hologram production, owing to the slowing of military contracts in the post-Soviet era. Over that decade, holography exhibitions, book publications, and dedicated conferences all declined twofold.

Institutional locales had different engagements with holograms. The Science Museum, London, mounted a permanent holography display at the entrance to its Optics Exhibit in the early 1980s (see figure 6.1). The exhibit followed the third successful London holography exhibition, Light Dimensions (originally mounted at the Royal Photographic Society in Bath and later at the Science Museum) which, with the earlier Light Fantastic exhibitions, amounted to the most concentrated and well-attended shows in a Western city. Providing a good range of holograms, the Science Museum exhibit had nevertheless deteriorated some-

FIGURE 6.1
Entrance to the Optics gallery of the Science Museum, London, 1988, showing a pulsed hologram of Dennis Gabor, inventor of the principle of holography. Courtesy Science Museum, Science & Society Picture Library.

what by the mid-2000s, with failed lighting arrangements and chemical deterioration of some holograms. The entire optics gallery—of which holograms had been the leading exhibit—was closed in January 2006, and holograms were relegated to the museum stores.

The cultural uses of holograms in a museum environment developed independently in the Soviet Union. There, both the type and purpose of holograms were differently conceived. While holograms in the West were generally of the Leith-Upatnieks type (requiring a laser light source) during the first decade, and subsequently a rainbow or rainbow-multiplex type from the mid 1970s, holograms in the Soviet Union were usually of the Denisyuk type, which reflect a green-yellow image from a white-light source. These Soviet holograms proved straightforward to record (although they were considerably more sensitive to vibration of the apparatus during exposure) and to illuminate (a bright light bulb or projector sufficed).

But the *purpose* of Soviet holograms eschewed aesthetic expression per se. Soviet holographers, exclusively scientists who had access to the expensive equipment, made holographic images either as part of their scientific research or to record precious objects. The latter application was well-suited to museums, because the priceless works of material culture in Moscow, Leningrad, and Kiev museums could faithfully be reproduced and displayed in provincial centers for the first time. Thus art holography in the Soviet Union was conceived as high-quality recording and popularization of art treasures, rather than the creation of a new form of optical art as in the West.

This art reproduction work had been begun by Gennadi Sobolev of the Cinema and Photographic Research Institute in Moscow (known by its Russian acronym NIKFI) during the late 1960s, when he and his colleagues recorded reflection holograms of museum objects from the Hermitage and Kremlin Museum collections. These toured the country and were later widely seen and admired internationally. Vladimir Markov and his colleagues took up similar work in the Ukraine from 1976, when a group from the Institute of Physics of the Ukrainian SSR Academy of Sciences and the Ministry of Culture gained privileged access to Kiev museums.[20] The large-scale and high-quality reflection holograms produced in Tbilisi, Leningrad, Moscow, and Kiev had a dramatic impact on Western observers.

By contrast, the MIT Museum in Cambridge, Massachusetts, had different reasons for staging a holography exhibit. MIT created the Media Laboratory in 1984, with one group of researchers led by Stephen Benton, formerly of Polaroid, and, from the 1970s, the chief spokesperson for American display holography. Benton joined MIT in 1982, where he pursued holographic imaging research. A second connection with holography was MIT's Center for Advanced Visual Studies (CAVS), founded in 1967 and which since the mid-1970s had supported holographic art through participants such as Harriet Casdin-Silver. CAVS also was a contributor to the master of arts in visual studies, which mixed art and technological themes.

After the New York Museum of Holography closed in 1992, Benton, MIT Museum director Warren Seamans, and MIT president Charles Vest (himself a former WRL engineer specializing in holographic interferometry) raised funds to buy the MoH archives and its hologram collection outright, and the collection was moved to the MIT Museum a year later. Founded in 1971, the museum occupies a former radio factory on the edge of the

MIT campus with a number of other MIT tenants, and preserves, displays, and collects artifacts in five disparate subject domains significant for the institution's history.[21] The holography collection includes some 1,500 holograms acquired from the MoH and covers a period from the mid-1960s to the late 1980s. Only a few dozen holograms can be displayed owing to space restrictions, and the MIT Museum attracts somewhat fewer visitors than did the New York Museum of Holography.[22]

Why Collect? Why Display?

Given the history of the subject and of its public representation, we may ask what role museums should play in preserving and elucidating holography. How (and, indeed, why) should the historical trajectory of holography be explained?

There are competing accounts and justifications: historians sometimes disagree with museum curators, who may disagree with practicing holographers. For practitioners, the subject had come to be represented in historical terms as early as 1971, when Dennis Gabor won the Nobel Prize for Physics. Historical accounts were essential to pursue the early priority claims that were crucial to the awarding of the prize and to resolving patent disputes. Subsequent accounts were less personally directed, but promoted the prevailing expectations of technological progress and the expansion of scientific knowledge. As late as the 2000s, however, many practitioners resisted the historicity of their subject, perhaps seeing this as an unfavorable way of judging its trajectory alongside its prognostications or of relegating it to the past. It is equally difficult for those still active in the field to recognize their activities in a historical sense.

But, of course, histories do inevitably get constructed, often without the direct intervention of the participants and frequently in a simplified form that serves particular agendas. Direct interaction of the historian with those practitioners is ambivalent: on the one hand, holographers provide direct (if occasionally conflicting) personal accounts and interpretations of episodes; on the other, they may resent the interference of an outsider seeking to explain events in ways that may not actively support their interpretations or promote the subject as they would themselves. These problems are exacerbated for commercial holographers, who understandably may want to portray their industry in a positive light. The historian's account may conflict with others that inevitably suffer from selective recollection and reshaping and rehearsing of events to satisfy simplified chronologies and accounts.

Just as important, popular understanding of the subject may conflict with both practitioners' and historians' accounts. Public portrayals of holograms since the 1960s emphasized their ability to shock and awe. Holograms entered the realm of the sublime through their connection with the viewing environment and natural optical phenomena. Consider the rainbow appearing after a severe storm and dividing the sky into bright and dark regions by its colorful arc; ice haloes appearing in the fog of a cold morning and noticed from an isolated hilltop; bright lightning bolts in an otherwise inky sky; the *heiligenschein*, a halo of light surrounding the shadow of one's head reflected from the dewy grass of morning; or the *glory*, a circular rainbow around the shadow of an airplane as it flies above a cloud. One reason for the impact of these optical phenomena is that each is experienced in a peculiar context that isolates the observer from the everyday world. The same transcendent experience can be captured in laboratory or gallery demonstrations of holograms. Like natural phenomena, the viewing experience surprised or awed observers because of the unfamiliar imagery and disorienting environment.

But while a sense of awe is evanescent, other popular evocations of holograms proved durable. The first science fiction short stories incorporating holograms appeared in the early 1970s.[23] Films, most significantly *Star Wars* (1977), soon developed the theme of fictional holograms as projected, often miniaturized and animated three-dimensional images.

Television science fiction began to employ holograms as plot devices soon after *Star Wars*, but with distinctly different attributes. The British comedy television series *Red Dwarf* (broadcast in Britain 1988–1994, and set onboard a spaceship of the distant future) extended the capabilities of fictional holograms far beyond contemporary expectations. One of the characters was a hologram, calculated by the spaceship's computer, his form reconstructed by a "light-bee," which rapidly painted his three-dimensional image in real time. In early episodes, the character was described as being generated by "Soft Light," and could not be touched; he was later said to be rewired with a "Hard Light drive" to make him solid, thus extending the notion of a hologram from a mere optical effect to one imbued with mechanical, as well as electronic, underpinnings. A similar idea was developed in the second *Star Trek* television series *The Next Generation* (broadcast 1987–1994) with its plots incorporating a "holodeck," or holographic visualization room, in which an entire environment can be reconstructed by computer (and, indeed, interacted with by the human characters). As in *Red Dwarf*, the fictional optical technology was gradually augmented by other capabilities: some objects in the holodeck were described as being "replicated" (that is, material objects manufactured from individual atoms) and animated with "weak tractor beams" and "shaped force fields." In the succeeding series, *Deep Space Nine* (1993–1999), the technology becomes commercialized: "holosuites," or holodecks of various capacities, are rented out for private use. And in the *Voyager* series (1995–2001) these technologies are used to implement an Emergency Medical Hologram, or virtual doctor. This virtual character, and his tribulations as a sentient computer program embodied in a computer-generated hologram, spawned a popular book.[24]

Thus the ontology of popular belief was extended: the hologram became a staple of science fiction plots, alongside time travel, robots, black holes, and interplanetary travel. Indeed, one online science fiction magazine adopted the title *Hologram Tales* not because of a central interest in these plot devices or their underlying technology, but merely because of the ubiquity of the concept in contemporary science fiction. Most important, science fiction holograms were paradigms of progress, like the tales in which they were embedded. From the late 1970s, then, there was an increasingly obvious bifurcation of technologists' and science fiction writers' and audiences' perceptions of the hologram.

Historians have their own cluster of interests and interpretations, situated somewhere between the technical concerns of practitioners and the more naive themes of the wider public. They, too, may look both backward and forward, trying to understand possible futures in terms of what has come before. Why, for example, has holography evolved in the way that it has? What shaped the directions it took? Can its history tell us anything about the nature of discovery, invention, marketing, popular engagement, or progress?

Just as holography allows observers to see an image with multiple perspectives, exploring its history demands a range of viewpoints. Its past has never been summarized adequately by its practitioners, because the subject has been divided up by scientist-engineers, artists, artisans, and entrepreneurs. Each group assessed the history, success, and future of its subject differently. The hundreds of capsule histories of holography written since the 1960s, appearing in newspapers, magazines, conference proceedings, scientific papers, introductions to books, and holographers' folklore, have been written at distinct times for distinct audiences, and they often came to dramatically different conclusions about which ideas, events, players, and products were important.

How can these different users' understandings of the subject be reconciled? The role of artifacts can be significant in embodying or reifying a sense of history. Hologram exhibitions frequently have been used to make the evolution of holography tangible. Nevertheless, the desire to locate missing links can misrepresent, too, as Emmett Leith reflected concerning the preservation of early holograms in a 2003 interview:

> People ask, "Well, what was your first hologram, which is the first hologram?" And museums around the country and private collections and so on, people have enough of the "very first hologram" around [like pieces of the real cross during the Middle Ages, when] you could find, in crypts and grottos, enough pieces of the real cross to start a lumber yard. And some of these holograms that people claim to be the original might be the thousandth.[25]

Attributing a relic-like identity to holograms deemed to be historically important began during the late 1970s, when a historical perspective was becoming established. The flurry of large public exhibitions and retrospectives during that period sought to chronicle a clear history of the young field. The Museum of Holography in New York, which organized some of the first large exhibitions, became, for a time, the repository for these significant objects. Religious parallels can be suggested: the identification of relics (carefully transported from one temporary place of veneration to another)—indeed, tales about transporting important holograms to exhibitions are recounted that assume the dimensions of pilgrimage journeys to shrines in Chaucer's *Canterbury Tales*; the multiplication of holographic relics as Leith describes, and their rapid escalation in value; their display in carefully oriented reliquaries; their home in a dark and respected sanctuary.

Such musings provoke the question of the purpose and future of historical collections. Museums and galleries actively construct popular history. With the perception of holograms as historical objects and a material culture to be preserved, a relationship grew between holograms, museum curators, and their representation of history.

However, the uneven preservation of the documentary and material culture of holography illustrates the peripheral status of the field in wider culture. The papers of Dennis Gabor were collected and archived by Imperial College largely because of the status he achieved late in life with the award of his Nobel Prize; another important early worker largely forgotten after the 1960s boom of the subject, Gordon Rogers, donated his own career files to the Science Museum. Even so, the curator at the New York MoH noted that Gabor's Nobel Prize was becoming mildewed owing to damp conditions there; when taken over by the MIT museum, the hologram collection went unarchived for some years, and again did not benefit from the environmentally controlled storage necessary to inhibit chemical reactions in some of the unstable emulsions; and some of Gordon Rogers' film records at Imperial College have congealed into a decaying mass.

The papers of few other early practitioners have been preserved. A survey of the holdings in Ann Arbor, the crucible of development of the subject academically, commercially, and artistically during the 1960s, shows that there are historically important documents, equipment, and holograms scattered around the small city, but no historical collections or exhibits focusing on them. The Bentley Historical Library of the University of Michigan in Ann Arbor holds some of the administrative records of Willow Run, for example, but does not identify holography as a specific collecting category, nor does it presently hold much archival material specifically on the subject. Most documents remain in the hands of individuals—participants as students, entrepreneurs, classified research workers, or commercial engineers—still living in the area.

Nor have surviving firms and institutions made more than a casual attempt to preserve their past, to say nothing of the majority of failed business enterprises. Carl Aleksoff, for instance, at Willow Run Labs as a student and then with its successors the Environmental Research Institute of Michigan (ERIM), ERIM International, Veridian, and General Dynamics, recalled that optical processing equipment reaching the end of its working life was revealed to visitors, but ultimately neglected:

> For a number of years these things were displayed in the lobby—but it's all been thrown away. You can only do it for so long. For a bottom line profit motive, you've got to account for how much area you have—so many square feet—what are you doing with it? How productive is it?[26]

Despite the relatively long-term stability of organizations funded principally by military contracts, WRL, ERIM, Veridian, and General Dynamics suffered from the demands of secrecy, which are incompatible with the preservation of open history. This was equally true of the more commercially oriented Ann Arbor holography firms such as the Conductron Corporation, KMS Industries, and GC-Optronics. These companies were simultaneously constrained by classified contracts and by the desire to control commercially useful proprietary knowledge, on the one hand, and the desire to vaunt technologies that they hoped would become major income-generating streams, on the other.

This patchy preservation is not restricted merely to commercial firms and classified-research organizations. The MIT Museum collection of holograms is disproportionately distributed, with more holograms from the mid-1970s to 1980s when the MoH was most active, and few after the mid-1980s when the MoH encountered more serious financial and administrative difficulties. The associated archives of the MoH held at the MIT Museum provide an excellent snapshot of holography's most fertile and expansive period as a cottage industry and would-be art form, 1975–1985. This leaves, though, a substantial period little represented in archival collections. Despite the commitment to preserve and make available these resources, the MIT Museum has not had the luxury of a permanent curator of holography, controlled-environment storage conditions for the collection, nor an explicit collection policy that enabled it to continue to acquire representative examples of holograms or documentary records.[27] This is perhaps understandable for a university museum that has a remit primarily to document the institution itself. However, MIT has been a major participant in holographic research through the Media Lab, so the holography collection arguably conforms to the Mission Statement, aiming to "document, interpret and communicate, to a diverse audience, the activities and achievements of the Massachusetts Institute of Technology and the worldwide impact of its innovation, particularly in the field of science and technology."[28] Interestingly, relatively few of the Media Lab holograms were displayed at the museum during Stephen Benton's life, although there was a limited exchange of examples serving as demonstration items for courses. A collecting policy was drafted in 2001,[29] but limited funding and curatorial resources restricted new acquisitions.

The MIT Museum has attempted consciously to make best use of its holography collection while serving other commitments. Its limited resources are not unusual, however. Larger national museums—the Smithsonian Museum in Washington, the Science Museum and Victoria & Albert Museum in London, the National Museum of Photography in Bradford, UK, and the Deutsches Museum in Munich, for example—each of which has mounted holographic displays, commissioned or acquired holograms—have not established collecting policies to preserve the ephemeral material culture of holography (Chris Titterington, while assistant curator of photographs at the Victoria & Albert Museum in London from the 1980s until 1995, established a hologram collection policy, but this did not outlive his tenure). This, too, may be understandable, if one assumes the remit of technology or cultural museums to be the recording and valorizing of technologies perceived to be successful, relevant, or influential. Historians increasingly question such asymmetrical representation of the past, however. It biases the historical record to suggest that progress is natural and straightforward and that subjects declining in economic or popular impact are unworthy of attention. As discussed above, a balanced treatment of perceived successes and failures is necessary not only to understand past events, but also to learn from them. Yet museums, defined by their sponsors, remits, and audiences, are often compelled to present stories of progress. By being pigeonholed in this way, the history of holography is pared down to an unfaithful representation. As a result, there has also been an understandable dissonance between the stories told for different audiences.

Holography for Museum Curators

As outlined above, the 1990s witnessed a retrenching of aspirations and a migration and reorganization of major hologram collections from private owners to institutional homes.

Nevertheless, this did not promote cultural or physical stability: the institutions were under-funded to adequately archive, preserve, and extend hologram collections, and for the most part they had no remit to collect materials beyond holograms themselves. Nor did major national institutions choose to collect holographic materials in a concerted way. In the UK, the National Museum of Photography in Bradford and the Victoria & Albert Museum and Science Museum in London all maintained the token collections acquired in the 1980s. The Smithsonian Museum in Washington, D.C., and the Greenfield Village Henry Ford Museum in Michigan acquired some historic materials from Ann Arbor holographers in the 1990s, but not as part of a concerted collections program. And university archives, notably the Bent-ley Historical Library in Ann Arbor, have also largely ignored the voluminous documentary materials still available from former employees of WRL, the University of Michigan engi-neering departments, and holography firms in the area.

There are at least two pragmatic reasons for this lack of preservation. First, holograms themselves are of uncertain physical stability. During the 1970s and 1980s, holographers sought to increase the diffraction efficiency, and brightness, of holograms by various bleaching processes, many of which have undetermined archival characteristics. Other holographic media, notably DCG, are adversely affected by humidity if edge seals break or deteriorate.

Second, holograms are difficult to display. All holograms, even "white-light" varieties, require a carefully chosen and oriented light source. Leith-Upatnieks holograms require a laser or laser diode; all others require a near point source of bright white light, such as a halo-gen bulb. Light sources must illuminate their holograms near the design angle to ensure a bright, undistorted image, which means that in practice each hologram requires separate illu-mination. And finally, most holograms have a relatively narrow viewing angle: although the imagery can sometimes be almost indistinguishable from the original object, the hologram usually must be viewed from within a "window" angle less extensive than would be expected in viewing the object itself. These display requirements mean that the viewing geometry must be recorded and reproduced faithfully by museum staff, and that relatively power-hungry light sources must be checked regularly. The lasers and halogen lamps employed in holo-gram displays until the late 1990s can now be replaced by laser diodes and LEDs, which have considerably lower cost, energy consumption, and deterioration with age, but this con-version has to date been pursued only sporadically.

Despite such problems, from a historian's point of view the case can be made that holog-raphy represents a fascinating and important case of modern science and technology. It is a complex example of a surprisingly common but little noticed situation in modern science, in which a technical subject has created new communities and grown with them. Its evolution has been distinctly different from what most holographers—and even historians of science—might have expected, which can help us to better understand how modern sciences emerge, and how to more realistically chart their future trajectories. And because of the rich variety of communities that the subject has embraced, ranging from artists to defense contractors, its history is likely to be of enduring interest to broad audiences.

The Potential for a Contentious Subject

Museum exhibits focusing on scientific controversy can be an effective way of informing the public about the resolution of intellectual disputes by experts. Exhibits focusing on contentious applications can provoke questions and discussion about the social purpose and ethics of science. Holography—with its history of gradual assimilation and of military origins—is one such relevant case. But, more cogently, holography is an example of technological evolution that challenges preconceptions.

Museums of science and technology have striven increasingly to portray more complex stories of development than their predecessors. No longer can technology itself be represented as an unalloyed good for society; no longer is technical progress itself seen as unproblematic. Holography exemplifies an intellectual concept that greatly generalized concepts of optics and imaging, and of a technology that, for a quarter century, excited considerable enthusiasm from corporate researchers, university scientists, artisans, and artists. This pulse of interest has weakened, as the inflated expectations of technical advance did not materialize. On the other hand, the subject has continued to rise in popular appeal via science fiction, where holograms continue to represent the future. I would argue that this recasting of the subject from scientific to technical to popular culture is a common process but one particularly well illustrated by holography, which straddled cultural domains.

Holography has a place in museum representations, but unlike the uncritical exhibits of a quarter century ago. It can illustrate the following:

- Notions of technological evolution—the subject is not a case of *technological determinism*: holography did not continue to expand technologically or alter culture inexorably.
- Notions of success, failure, and progress—since the late 1960s these concepts in relation to technology have been increasingly problematized by both scholars and the general public. Holography is a case that is both appealingly wide-ranging and yet largely uninfluenced by strong cultural polarization (unlike, for example, nuclear energy or human cloning).
- Representations of new technologies in terms of familiar analogies—holography was widely depicted as a form of photography during the 1960s, a categorization that both underrepresented its novel features and associated unrealistic expectations with it (notably in the form of holographic television and outdoor imaging).
- Appropriation of technologies by disparate user communities—the "user groups" of holography included not only scientific, artisanal, and artist holographers, but also manufacturers and science fiction enthusiasts, all of which had varying degrees of interaction.

The development of museum exhibits exploring these themes is a challenge, but the expense of building museum collections demands even more justification and energy. The preservation of a historical record that moved from classified research to countercultural groups to popular culture would demand considerable resources. Nevertheless, the potential of such collections and archives for scholars and for museum curators—exploring innovation, aesthetics, creativity, the role of social groups, and the application of scientific knowledge—is profound. And, most important, the opportunity to acquire such materials is time limited: while the materials and documents of recently deceased well-known holographers such as Stephen Benton (d. 2003) and Emmett Leith (d. 2005) are very likely to be collected and preserved, those of their lesser-known contemporaries may not be. As the Malian writer Amadou Hampâté Bâ (1901–1991) said of his continent's oral history at a 1962 UNESCO meeting, "Quand un vieillard meurt, c'est une bibliothèque qui brûle" (When an old person dies, a library burns). To fail to preserve such information would be collectively to forget a significant, and ongoing, technology that has directed the lives and careers of several thousand persons and represented late-twentieth-century aspirations concerning the future for many more.

Acknowledgments

This work was supported in part by grants from the British Academy, Carnegie Trust, Royal Society, American Institute of Physics' Friends of the Center for the History of Physics, and Shearwater Foundation.

Notes

1. Johnston, *Holographic Visions*.
2. For example, Lippmann, "La photographie," 274. For accounts by scientists stressing these intellectual roots, see Kirkpatrick, "History of Holography," and Scanlon, "Holography: Simple Physical Account."
3. Zernike, "How I Discovered Phase Contrast."
4. Johnston, "From White Elephant." The Gabor papers are held at Imperial College Archives, London, and a smaller collection of Gabor's papers from his consultancy at CBS Laboratories is held at MIT Museum in Cambridge, Massachusetts.
5. See, for example, Denisyuk, "On the Reflection."
6. Johnston, "Absorbing New Technologies."
7. Johnston, "Telling Tales."
8. Johnston, "Explosion With a Slow-Burning Fuse." A number of the oral interviews between the author and Ann Arbor workers, including Leith (now deceased) and other members of WRL and commercial firms, have been deposited at the American Institute of Physics, Neils Bohr Library, College Park, Maryland.
9. See, for example, Ennos, "Holographic Techniques"; and the first international conference to the subject held in Glasgow in 1968, Robertson and Harvey, *The Engineering Uses of Holography*.
10. Johnston, "Attributions of Scientific and Technical Progress."
11. Charnetski, "Impact of Holography."
12. Accounts of the group are divergent; relevant sources include Anon., "School of Holography Flourishes on West Coast"; Cross, "The Story of Multiplex," 6; Gorglione, "Lloyd Cross."
13. See, for example, Bohm, *Wholeness and the Implicate Order* for Bohm's identification of holographic characteristics of the physical universe; and Goleman and Pribram, "Holographic Memory,"

for a popular account of Pribram's speculations on holographic properties of the mind, in particular the lack of localized memories in the brain.

14. Johnston, "Shifting Perspectives."

15. Jackson to Benton, May 19, 1978.

16. The quoted details of the New York Museum of holography are based on extensive archives purchased after its bankruptcy in 1992 and now held at MIT Museum, Cambridge, Massachusetts.

17. Lancaster, "The Museum of Holography."

18. Misselbeck, "The Museum für Holographie."

19. Wood to Tomko, June 13, 1991.

20. On the Russian-Ukrainian perspective on the application of museum holograms, see Yavtushenko and Markov, "Holography Serves Ukrainian Museums"; Markov and Mironyuk, "Holography in Museums of the Ukraine"; Markov, "Holography in Museums"; and Markov, "Display and Applied Holography."

21. Besides holography, the MIT Museum collection categories include architecture, nautical design, science and technology, and MIT ephemera (Connors, Benton, and Seamans, "Report from the MIT Museum").

22. When the holography collection was acquired, the museum opened an exhibition of holograms in the main gallery in 1994 and attendance figures increased dramatically. Nevertheless, the gallery space was later subdivided, with the hologram exhibit reduced and moved behind a more popular exhibit on artificial intelligence.

23. Bryant, "The Poet in the Hologram"; Gibson, "Fragments of a Hologram Rose."

24. Picardo, *The Hologram's Handbook.*

25. Leith to Johnston, interview.

26. Aleksoff to Johnston, interview.

27. The conditions were nevertheless a significant improvement over the MoH in New York; see Dinsmore, "On MoH Collection."

28. Anonymous, "Strategic Plan."

29. Anonymous, "DRAFT MIT." Following appointment of a new director in 2005, a post of permanent curator of holography was mooted but remained unconfirmed in 2008.

References

Aleksoff, Carl to S. F. Johnston. Interview, September 9, 2003, Ann Arbor, S. F. Johnston Collection.

Anonymous. "DRAFT MIT Museum Collecting Guidelines for Holography." MIT, August 2001.

Anonymous. "School of Holography Flourishes on West Coast." *Holosphere* 2 (1973): 1, 5–6.

Anonymous. "Strategic Plan, November 2000–June 2005." Report, MIT, 2000.

Bohm, David. *Wholeness and the Implicate Order.* London: Routledge, 1980.

Bryant, Edward. "The Poet in the Hologram in the Middle of Prime Time." In *Nova 2: Original Science Fiction Stories*, edited by H. Harrison116–32. London, 1972.

Charnetski, Clark. "The Impact of Holography on the Consumer." Presented at Electronics and Aerospace Systems Conference Convention. Washington, DC, 1970.

Connors, B. A., S. A. Benton, and W. Seamans. "Report from the MIT Museum." *Fifth International Symposium on Display Holography* 2333 (1994): 146–51.

Cross, Lloyd. "The Story of Multiplex." Transcription from audio recording, Spring 1976. Ambjörn Naeve Collection. http://kmr.nada.kth.se/wiki/Amb/MyHolographyArchive (accessed 15 September 2007).

Denisyuk, Yu N. "On the Reflection of Optical Properties of an Object in the Wave Field of Light Scattered by It." *Doklady Akademii Nauk SSSR* 144 (1962): 1275–78.

Dinsmore, Sydney. "On MoH Collection from Curator." Report, MIT Museum 10/192, October 18, 1990, and January 22, 1991.

Ennos, A. E. "Holographic Techniques in Engineering Metrology." *Proceedings of the Institution of Mechanical Engineers* 183 (1969): 5–12.

Gibson, William. "Fragments of a Hologram Rose." In *Burning Chrome*, edited by W. Gibson. New York: 1977. http://lib.ru/GIBSON/frag_rose.txt (accessed 30 July 2006).

Goleman, Daniel, and Karl Pribram. "Holographic Memory." *Psychology Today* 13 (February 1979): 71–84.

Gorglione, Nancy. "Lloyd Cross." *Holographics International* 1 (1987): 17, 29.

Jackson, Rosemary H. to S. A. Benton, May 19, 1978. New York. MIT Museum 26/584.

Johnston, Sean F. "Telling Tales: George Stroke and the Historiography of Holography." *History and Technology* 20 (2004): 29–51.

Johnston, Sean F. "Attributions of Scientific and Technical Progress: The Case of Holography." *History and Technology* 21 (2005): 367–92.

Johnston, Sean F. "From White Elephant to Nobel Prize: Dennis Gabor's Wavefront Reconstruction." *Historical Studies in the Physical and Biological Sciences* 36 (2005): 35–70.

Johnston, Sean F. "Shifting Perspectives: Holography and the Emergence of Technical Communities." *Technology & Culture* 46 (2005): 77–103.

Johnston, Sean F. "Absorbing New Technologies: Holography as an Analog of Photography." *Physics in Perspective* 8 (2006): 164–88.

Johnston, Sean F. "Explosion with a Slow-Burning Fuse: Origins of Holography in Ann Arbor, Michigan." *Proceedings of SPIE* 6252 (2006): 1–15.

Johnston, Sean F. *Holographic Visions: A History of New Science.* Oxford: Oxford University Press, 2006.

Kirkpatrick, Paul. "History of Holography." *Proceedings of the SPIE, Holography* 15 (1968): 9–12.

Lancaster, I. M. "The Museum of Holography—Its Role and Policies in a Changing Environment." Presented at Practical Holography II. Los Angeles, CA, 1987.

Leith, Emmett to S. F. Johnston. Interview, January 22, 2003. Santa Clara, CA, S. F. Johnston collection, University of Glasgow in Dumfries, Rutherford-McCowan Building, Crichton University Campus.

Lippmann, Gabriel. "La photographie des couleurs." *Comptes Rendus de L'Academie des Sciences* 112 (1891): 274.

Markov, V. B. "Holography in Museums—Why Not Go 3-D." *Museum* 44 (1992): 83–86.

Markov, V. B. "Display and Applied Holography in Museum Practice." *Optics and Laser Technology* 28 (1996): 319–25.

Markov, V. B., and G. I. Mironyuk. "Holography in Museums of the Ukraine." Presented at Three-Dimensional Holography: Science, Culture, Education. Kiev, USSR, 1989.

Misselbeck, Reinhold. "The Museum für Holographie und neue visuelle Medien and Its Influence on Holography in Germany." *Leonardo* 25 (1992): 457–58.

Picardo, Robert. *The Hologram's Handbook*. New York and London: Simon & Schuster, 2002.

Robertson, Elliot R., and James M. Harvey. *The Engineering Uses of Holography*. Cambridge: Cambridge University Press, 1970.

Scanlon, M. J. B. "Holography: A Simple Physical Account." *GEC Review* 8 (1992): 47–57.

Wood, Glenn P. to M. Tomko. June 13, 1991. MIT Museum 2/38.

Yavtushenko, I. G., and V. B. Markov. "Holography Serves Ukrainian Museums." *Museum* 34 (1982): 168–71.

Zernike, Frits. "How I Discovered Phase Contrast." Nobel Lecture, Nobel Prize for Physics (1953); also *Science* 121 (1955), 345–49. www.nobelprize.org/zernike-lecture.pdf (accessed 7 June 2005).

Inside the Atom

TWO SIDES OF A STORY

Jane Wess

THIS CHAPTER concerns the rationale behind a small exhibit called Inside the Atom, which opened at the Science Museum, London, in early 2005.

In 2002 and 2003, the Science Museum undertook a large-scale restructuring and a reassessment of the work of curators. The curatorial resource was divided into four areas: science, medicine, information and communications technology, and engineering. Each group developed an interpretation strategy. The science curatorial group drew up a strategy that included questioning the nature of science and the impact of science. Within the nature of science strand, the group embarked on an exploration of the concept of *scientific breakthroughs*. Alongside this interpretation strategy, the group, along with other curators, developed a framework for a new science gallery, a large project to open in 2013. The breakthrough strand became an element of this bigger project, so the small exhibit had an advantage in that it was seen as a trial run for the larger gallery.

Because breakthroughs formed part of the overall strategy did not mean the test project was approved: there were no funds available, and the museum's focus was on large-scale projects. The trigger for the go-ahead was the completion of an acquisition, whereby the museum acquired the full set of medals from the Thomson family. We now have J. J. and G. P. Thomson's Nobel Prize medals, Royal Medals of the Royal Society, and J. J. Thomson's Copley medal, along with a number of less well-known ones. The Institution of Electrical Engineers (IEE) (now the Institution of Engineering and Technology) kindly donated money toward their display; with some internal funding gleaned from other projects, there was a viable exhibition budget. Initially it was intended to show the medals as new acquisitions, but this topic had the potential to take on the challenging role as pioneering the strategy to explore breakthroughs. After all, nothing illustrates the history of science as a series of "great men" better than a Nobel Prize medal, so this acquisition was particularly timely.

The museum has a large showcase on the second floor that had previously been used to display cutting-edge science and technology. However, the Wellcome Wing, which opened in 2000, made a large space available for contemporary science, thus releasing this case for more diverse outputs. The case is very tall, quite long, but not particularly wide. The physical possibility of showing literally two sides of a story was irresistible. The story begging for attention was J. J. Thomson's controversial "discovery" of the electron, but it was decided to widen it to make a general point about breakthroughs and to exhibit more of the collections. The exhibition therefore focuses on the fairly homogeneous and unbroken chain of research into electricity in gases at low pressure. Inside the Atom: Two Sides of a Story is a history of a type of scientific apparatus as much as the history of a train of thought, although they are closely connected. This is a presentation that plays to the strengths of museum exhibitions as opposed to a written account. The exhibit was intended for students ages fourteen and up, an unusually sophisticated target group for the Science Museum at this time.

The concept of a breakthrough has a particular resonance for museums because, in keeping with many popular historical texts on science, we have tended to bolster the Smilesian "great men" stories with our counterparts the prototype "firsts," and our "iconic objects." Just as historians of popular science have to have a rationale for choosing certain stories above others in terms of greatness and importance, museums have had to choose particular objects above others, leading to the conversion of objects into relics. As well as being a feature of older galleries, Making the Modern World, which opened in 2000, has a row of iconic objects running down the center, so this is a persistent method of approach which is used to attract and impress visitors. Making the Modern World also contains everyday objects arranged according to a classification system of the time, so there is a balance, with objects playing different roles. However, nowhere in the museum was there anything to question directly the great men–great exhibits until Inside the Atom: Two Sides of a Story.

Serious studies in the history of science, using primary written sources, can put the claims for breakthroughs made on behalf of the great men within a context of comparable greatness and richness of approach.[1] Similarly, in a museum environment we can find, if we look at our iconic objects under the microscope, that their iconicity melts away, without making them any less interesting to a historian. Perhaps they are one of a series of evolving instruments, which only gradually become defined as important, perhaps earlier similar devices can be found, perhaps they have been modified so their original use is not clear. A case in point is the iconization of the e/m tube in the Science Museum, once used by J. J. Thomson, which was studied by Alan Morton. Morton examined how it was described at various points in its career.[2] In an effort to add value to the object, the claims made for its uniqueness and importance were inflated as time went on. Iconization is a feature of the historiography of science, and as such it is vital that we understand and recognize it. Iconization can happen immediately as an experiment or discovery is made, or it can be delayed, but either way, it is a social construct, and either way, it is part of the history of science. A few scientists, such as Wilhelm Conrad Röntgen, become famous almost immediately, while others, such as Albert Einstein, did not become household names until many years after their theories were published.

The idea behind the exhibition was to feature the work of three "great men," J. J. Thomson, Albert Einstein, and G. P. Thomson, on one side of the case, and to show all the other contributions to the field possible in the space allowed on the other (figures 7.1 and 7.2). Alongside the Thomsons, chosen for reasons already explained, Einstein was an obvious choice given that the photoelectric effect is a related topic, and the exhibition contributed to our offer in his anniversary year. The choice of subject fitted our intentions perfectly as J. J.'s breakthrough has come under some close scrutiny in recent years, in part due to the centenary of the electron in 1997.[3]

The general point of the exhibition was not to detract from the achievements of the "great men," but to emphasize that their work would not have been possible without the contributions of many others and that breakthroughs tend to happen in a context peopled by other scientists working along very similar lines. Charles Molan puts this very well: "Science history can over-emphasize the geniuses who make the vital intellectual link which represents a major step forward in our understanding of our universe. But these geniuses take their intellectual food from others who have planted and nurtured seed corn."[4] Also, that while these great men used particular pieces of equipment, very similar equipment was also being used elsewhere, and the essential work of those who designed and made the equipment must also be recognized. We wanted to bring out the communal aspect of the development of science in this period and, crucially for an exhibition, the communal nature of the technology.

The three great men were lined up on "three pillars of wisdom" in chronological order. They each have a title, portrait, and in the case of the Thomsons, the Nobel Prize medals and their iconic object; for J. J. the e/m tube and for G. P. the electron diffraction camera. In the case of Einstein, where instruments are conspicuously difficult to come by, a bust by Epstein was used. The text relates their principal discoveries, being careful not to overstate their cases. The Great Men side could be clean, uncluttered, and clear with the message that "he did this, with this, and got this in recognition." It was explained in as simple terms as possible what the apparatus did and what these men discovered in order to align the approach with that of the standard time-line account.

The other side of the case required much more work, and looks—quite deliberately—much more messy. It was cluttered, busy, vibrant. Less emphasis was placed on explaining the science with more on the social aspects, the showmanship involved, the communication networks, and the reliance on shared technology. It was decided to start from Jean Picard's chance discovery of the appearance of light in a barometer tube in 1676, and trace the thinking behind the observations concerning the phenomenon of electricity in low pressure gases from then on. Per F. Dahl's book, *A Flash of the Cathode Rays*,"[5] was a major source of technical information, and the subject scope of the exhibition is similar if the approach is somewhat different. There are no labels in this side of the case but the artifacts are numbered, and a booklet with a paragraph about each is fixed to the case wall. There is also a paragraph for each section.

FIGURE 7. 1
The "Great Men" side of "Inside the Atom: Two Sides of a Story" exhibition case, 2005. Courtesy Science Museum, London.

Eight sections were chosen ranging from First Flickers, describing Picard's observation, to the ultimate Inside the Atom, which showed the culmination of this particular line of interrogation in the opening up of the sub-atomic world. From the beginning, the importance of spectacle is stressed. In the first section there are no objects directly related to experiment in this area, only a seventeenth-century barometer by Daniel Quare, the earliest available in our collections. This highlights the concept that a rich seam of discovery was prompted by a chance observation relating to another branch of investigation. The early domestic barometer also chimes with some of the later demonstration items that were enjoyed privately by small gatherings and in Victorian drawing rooms.

The second section concerns the dissemination of scientific knowledge, focusing in particular on the prolific output of Francis Hawksbee on the topic of electricity in a rarefied environment. His performances at the Royal Society are represented by a replica of his "shower of fire," when drops of mercury were allowed to fall on a glass dome in air at low pressure. Fortunately the Science Museum had a replica made in the 1920s. Also displayed and based on his designs are the amber beads made for George III in 1761 to illustrate the light produced by electricity in a vacuum. The amber beads are attached to a wheel that was spun so the balls rubbed a flannel pad inside an exhausted chamber.[6] Other exhibits are an elegant late eighteenth-century table air pump by Dollond (figure 7.3), a cylinder electrical

FIGURE 7.2
The other side of "Inside the Atom: Two Sides of a Story" exhibition case, 2005. Courtesy Science Museum, London.

machine with Leyden jar, and an aurora tube (figure 7.4). The largely forgotten itinerant lecturers of this period relied on these routine pieces to create the spectacle necessary to attract an audience.

The third section looks at the growing desire to measure effects and to experiment to gain new knowledge, rather than simply demonstrate. The undeniably heroic Humphry Davy and Michael Faraday are joined by those less well-known, at least in this country, such as Martin Van Marum and Abbe Nollet. Van Marum is mentioned because of his experiments with different gases to produce different colors of discharge,[7] Nollet because he introduced an electric egg into his spectacular range of electrostatic experiments.[8] Faraday pulled these and other researches together in a systematic approach to distance of discharge, pressure, and type of gas, in a series of experiments undertaken in early 1838.[9] The discovery of the Faraday dark space is mentioned but not explained, not only because this would be challenging, but because explanations have changed over time, and the historical integrity of the exhibition was paramount. An early nineteenth-century electric egg, with moveable electrodes and a discharge

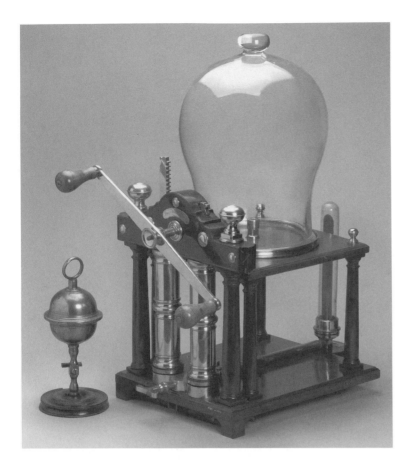

FIGURE 7.3
Table air pump by Dollond,
late eighteenth century.
Courtesy Science Museum,
Science & Society Picture
Library.

FIGURE 7.4
Cylinder electrical machine, late eighteenth century. Courtesy Science Museum, Science & Society Picture Library.

tube with connecting pump to show the supporting technology, serves to illustrate what was available before the technological developments of the mid-century (figure 7.5).

The fourth section focuses on the technological achievements of Heinrich Geissler and Heinrich Ruhmkorff (figure 7.6). The apparatus is no longer restricted to elegant brass and glass, but relies on substantial hardware. Geissler's beautiful tubes are shown to be dependant on his new revolutionary design of pump and on Ruhmkorff's improvements to the induction coil, which made it the staple electricity provider for scientists investigating this phenomena for the next seventy years.[10] Neither the pumps nor the induction coils are particularly

FIGURE 7.5
Electric egg by William Ladd, ca. 1860. Courtesy Royal Institution, Science & Society Picture Library.

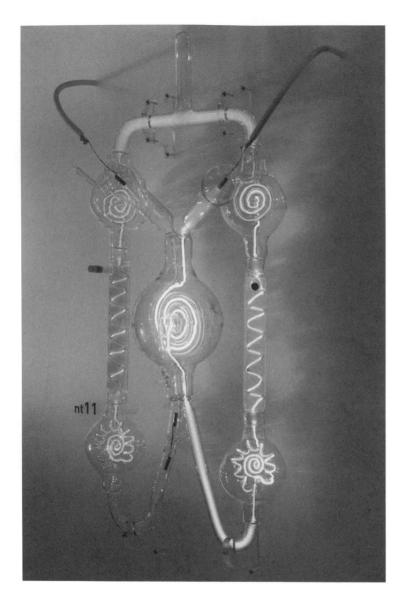

FIGURE 7.6
Replica by Jim Thomas
(2004) of tube made by
Heinrich Geissler in 1876.
Courtesy Science Museum,
London.

attractive, but they are as essential as the stunning tubes. Background equipment, which never features as iconic, as well as background experimenters who never get the limelight, are piled into the case.

Fortunately the museum has several original Geissler tubes. One is a relatively early example from 1860 with a relatively straightforward ray path. However, the spectacular element is provided by a working replica of a tube that Geissler sent to the special Loan Exhibition in South Kensington in 1876. The original was unfortunately too radioactive to exhibit under new legislation because of the presence of uranium glass. This tube exploits all the possibilities then available: colored glass, colored liquid, tubes within tubes, spirals, bulbs, and so forth. Respect for the craftsmanship of Geissler in making the tube was forced upon the exhibition team by the difficulties of the glassblower, Jim Thomas, who was commissioned to make the piece. Thomas had particular trouble in securing seals in the tubes within tubes. Eventually, thanks to his considerable skill and experience, this was achieved and the power is switched on for five-second bursts every ten seconds, thus giving a feeling of breathing, and is more effective in attracting attention than a permanently switched-on tube in an era when fluorescent tubes and neon signs are omnipresent. Tubes containing fluorescent minerals, related to the development of these light sources, are exhibited nearby.

The rather complex discovery of cathode rays is covered in the fifth section. The distinction is made between these and the colored streamers of light seen in less highly evacuated tubes, confusingly also often loosely referred to as cathode rays. Fortunately it was possible to find a distinctive quote by William Crookes. Cathode rays "absolutely refuse to turn a corner" while the streamers of light "will follow a tube through any amount of curves and angles."[11] It would have been useful here to have had a working demonstration of a tube being evacuated to see the changes, as this would have been helpful in allowing visitors to understand the issues facing the original experimenters and would not have been ahistorical. However, budgets and location did not allow us to do this. The section focuses on the properties of the cathode ray as observed by Julius Plücker, Johann Hittorf, William Crookes, and Philipp Lenard.

If William Crookes is not exactly unknown, he was not a Nobel Prize winner, and he never managed to attach his name to a breakthrough in the classic telling of the history of science. The other three are not well known in this country outside specialist circles. Julius Plücker, a very early observer of true cathode rays who employed Geissler as his technician, appears to have been the first to bend them in a magnetic field, the essence of J. J. Thomson's 1897 experiment.[12] Hittorf was responsible for the use of the Maltese cross shape as an obstacle in his experiments on the rectilinear propagation of cathode rays, not only proving his conjecture but giving rise to the ubiquitous piece of textbook apparatus.[13] With the tubes made by his technician Charles Gimmingham, Crookes was able to reach very low pressures. In view of this, and the new experiments that these tubes could perform, the next generation of tubes available were called Crookes tubes.

Lenard was a Nobel Prize winner in 1905, but while his experiments underpinned the discovery of X-rays and the discovery of the photoelectric effect, he did not manage to secure the recognition at that time that he thought he deserved. For a variety of reasons, some of them political as he became a strong supporter of the Nazis, his name is not among the great men of science. One point very pertinent to the theme of the exhibition is that his work on cathode rays depended on precious aluminium foil, which he sent to Röntgen, thus showing that ideas rely on experiments, which in turn rely on artifacts, and dissemination of all was taking place.[14] That science is a communal undertaking is brought out clearly here. The objects in this section include a Maltese cross tube, an original V-shaped tube belonging to William Crookes to show that true cathode rays travel in straight lines, and an educational Lenard tube from 1914.

The sixth section focuses on some of the people involved in popularizing this area of science, portraying them as colorful characters performing colorful experiments.[15] The work of Peter Gassiot, a wealthy wine merchant, Cromwell Varley, and William Crookes is featured in particular; Arthur Schuster and Johnstone Stoney are also mentioned. The museum is fortunate enough to have a pump used by Peter Gassiot, one of his spectacular revolving stars, and another demonstration designed by him: "Gassiot's cascade" made by William Ladd in the 1870s, which has similarities to Hawksbee's "shower of fire" of nearly 200 years earlier. William Crookes' electric radiometer shows how his interests in electricity in a vacuum fitted into the broader spectrum of his interest in radiation.[16]

While with hindsight the discovery of X-rays was an intellectual diversion from the understanding of what lay inside the atom, historically an exhibition on electricity in low pressures could hardly omit them as they had an impact on the research programs of the time, and from a technical point of view they formed a continuum in the experimentation using vacuum tubes. The penultimate section therefore focuses on X-rays and the extraordinary impact they had, actually setting back research on cathode rays for a year. We have an early X-ray tube exhibited by Alan Alexander Campbell Swinton to the Royal Photographic Society in February 1896, one of the well-known X-ray photographs of a hand with a ring dating from the same year, and a diagram of an early X-ray set up that includes a camera and induction coil. The latter is useful in this context because it shows the background apparatus necessary for the process.

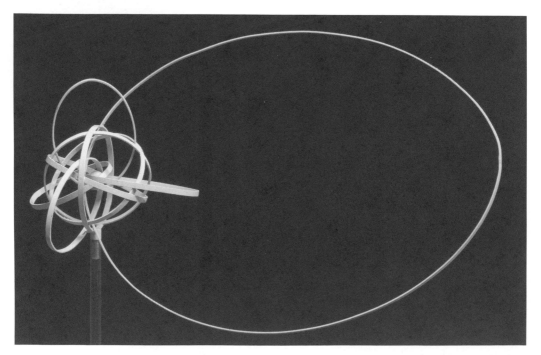

FIGURE 7.7
Model of sodium atom according to the Rumford-Bohr theory of 1911, donated by Lawrence Bragg in 1926. Courtesy Science Museum, Science & Society Picture Library.

Finally we reach the section, Inside the Atom and the year 1897, when the focus of activity returned to cathode rays. The findings of Emil Wiechert and Walter Kauffman are described in order to show how others were on the trail and that the suggestions and experimental findings of these scientists were comparable to those of Thomson. The e/m calculation was only one measure of the electron, a term not used by Thomson anyway, but the suggestion that electrons could exist inside the atom, whether it came directly from Thomson or was suggested to him, did open up a new field of physics. The models of the nuclear atoms of simpler elements, donated to the museum by Lawrence Bragg in 1926 (figure 7.7), represent the next step in the ongoing challenge to understand atomic structure. As these models are contemporary with G. P. Thomson's work on the other side of the case, and because the essential apparatus began to shift away from the evacuated tube technology, this seemed like a good place to stop.

On the Great Men side of the case there is a computer information point to take various aspects further. This enables a somewhat more discursive treatment and an increase in pictorial material. For example, the use of children and servants in electrostatic experiments by philosophers in the eighteenth century can be illustrated, as can the audiences for Faraday's lectures a century later. The various dark spaces that appear as the pressure is lowered can be shown through illustrations from contemporary textbooks. One story here is that of the skill of the instrument makers: Unsung Heroes. While some of these are featured in the exhibition, such as Hawksbee and Geissler, others, such as Peter Dollond, a prominent eighteenth-century maker; John Newman, who made instruments for Faraday; and Charles Gimmingham, who made Crookes' instruments, get a share of the credit. A lovely quote from Gimmingham's diaries record the frustration he felt when things went wrong, highlighting the crucial part he was playing in the process: "All went well till putting in bulb when all came to grief . . . this sort of thing makes me wretched."[17] Apart from the pages for historical characters there is also one for Jim Thomas who made the replica Geissler tube.

Another story is titled "Standing on the Shoulders of Giants." While it recognizes that Newton was probably only poking fun at Robert Hooke by using this phrase, the sentiment has a resonance with this exhibition. The strapline for this story is: "There are many instances in the history of science where scientists have benefited by the work of others and this work is not always remembered." Here the exhibition looks in more detail at the work of Johnstone Stoney, Julius Plucker, Philipp Lenard, Jean-Baptiste Perrin, and Max Planck, with a page each on their respective contributions.

Has the exhibition been effective? For small exhibitions the museum is unable to budget for audience evaluation so anecdotal evidence will have to suffice. When this chapter was delivered in spoken form in 2004, the exhibition had not yet opened and we were asking: Will we succeed in getting the message across? Do people need heroes? Will we be able to explain enough of the science to make people feel they have learned something? Can we use difficult science to make difficult points about science? Will people expect to be taught the history of science unquestioningly? It is pertinent to ask those questions again.

The exhibition is attractive, informed, uses our collections, and covers a "serious" science subject while we pursue funding for our much larger science gallery. Historians of science and educated adults understand the overall message. Some visitors will read one side or the other, but either is valid alone, so while this misses the strategic point it is not without benefit. The computer-information points appear to be used thoroughly when they are used, although they are used by only about a third of those who look at the exhibition. Possibly the sections do not come out very clearly, and it is only by reading the booklet that a good understanding of the science can be gleaned. So did we succeed in getting the message across? Yes, if by that we mean the message that science is a communal activity—this was understood by our target audience if they spent any time with the exhibition.

Do people need heroes? They come to the museum expecting to see something about Einstein at least, and many photographs have been taken of the Einstein column in particular by visitors, which acts as an attractor for the exhibition as a whole. From there people do appear to move to the lesser-known figures and the general story. The answer to the question is probably "yes." Audiences appear to need the hook of the known to draw them into the unknown, and a hero is a good start. In a culture where visitors have been taught history through heroes, and still regard the history of science in that way, it is easier to arouse interest through this viewpoint and then start to question it.

Will visitors have learned something? Certainly the people who use the CIP and the booklet will have learned something, and those who only look at the Great Men will be acquainted with their achievements. Even a glance at the exhibition will reveal the beauty of the instruments. Can we use difficult science to make difficult points about science? Yes, for an educated audience. The overall message is easier to comprehend than the physics, so it is quite possible to skip the latter and still get the main point that many people contribute to discoveries in different ways. We must be aware of the gulf between the level of curators of exhibitions and the level of the audience where science is concerned. Do people expect to be taught the history of science unquestioningly in the museum? They do appear to be very comfortable digesting straightforward pieces of information, but this may be changing with our new emphasis on debate.

To return to the more philosophical issue of the breakthrough: we need to understand the motives for its promotion. Identifying phenomena is a basic and omnipresent one affecting many school pupils' view of how science progresses. In science education names are attached to certain effects or laws, nearly always just one name, and so an individual is associated with a particular discovery, thus breeding the great man–great discovery view of the history of science from the start. Selling science is another, important for the government's agenda at the moment, and one in which national museums are encouraged to take part. The history of science has been populated by geniuses who make extraordinary discoveries, and their apparatus is wheeled out to promote their heroic status. Nationalistic competition is another, particularly rife in histories written before the more politically correct times following the end of Empire. Writers from areas considered peripheral justifiably feel a need to redress the balance.[18]

A fourth reason is simplification, and this is the most pertinent and pernicious to a museum setting, or anyone trying to write for a lay audience. Simplification versus a richer interpretation appears to be an essential tension of exhibition craft. In any other exhibition it would not have been considered good form to include so many characters, so many discoveries, or even so many artifacts. However, museum exhibitions do have a potential, far greater than the written form, to impart the richer picture to a general audience.

What the breakthrough view misses, as it focuses on a few great men having a few great moments, is the messiness of science. What a museum exhibition can do, far more easily than a popular book, is show this messiness without going into a level of detail that would swamp its audience. The beauty of an exhibition is that it can give a flavor of what is going on without asking the viewer to wade through masses of text. Unfortunately the exhibition style in the early twenty-first century mitigates against this with its insistence on clean lines, open spaces, and uncluttered cases. The educational function, such an important feature of modern museums, also requires simple messages. However, in a small exhibition such as this, there is more scope for experiment than in a larger scale gallery and more willingness to try something different. A desire to be intellectually rigorous and to tell new stories can only increase our chances of doing exhibitions that challenge received wisdom and open the eyes of our visitors to the workings of science in new ways.

Notes

1. The science group identified a lack of scholarly and semi-popular literature on the concept of breakthroughs. After the pioneering work of Paul Forman in "The Discovery of the Diffraction of X Rays by Crystals," there has been little until John Waller's *Fabulous Science*. Also of value is Brannigan, *The Social Basis of Scientific Discoveries*, with its emphasis on folk science and Golinksi, *Making Natural Knowledge*, especially the Coda, "The Obligation of Narrative."

2. Morton, "The Electron Discovered."

3. Arabatzis, "Rethinking the 'Discovery' of the Electron." See also Robotti, "J. J. Thomson"; Weinberg, "The First Elementary Particle"; Kragh, "J. J. Thomson"; Chown, "Just Who did Discover the Electron?"

4. Molan, "Who Discovered the Electron?"

5. Dahl, *Flash of the Cathode Rays*.

6. Morton and Wess, "Public and Private Science"

7. Chalmers, "Conduction of Electricity."

8. Nollet, *Leçons De Physique Experimentale*, vol. 6, leçon 20, plate 1, fig. 2, p. 249.

9. Faraday, *Experimental Researches*, esp. 484.

10. Hughes, *Networks of Power*, 86.

11. Crookes, "On the Illumination of Lines of Electrical Pressure."

12. Plücker, "On the Action of the Magnet."

13. Extract from Hittorf, "Ueber die Elektricitätsleitung der Gase." Hittorf's work is discussed extensively in Müller, *Gasentladungsforschung im 19. Jahrhundert*.

14. Letter to Röntgen, May 7, 1894. Quoted in Dahl, *Flash of the Cathode Rays*, 90, from Etter, "Some Historical Data."

15. See, for example, Noakes, "Telegraphy."

16. DeKosky, "William Crookes."

17. Fournier D'Albe, *The Life of Sir William Crookes*. Letters from Crookes to Gimmingham are in the Science Museum Library (MS 409) and will be available from late 2007.

18. Molan, "Who Discovered the Electron?"

References

Arabatzis, Theodore. "Rethinking the 'Discovery' of the Electron." *Studies in History and Philosophy of Modern Physics* 27 (1996): 405–35.

Brannigan, Augustine. *The Social Basis of Scientific Discoveries.* New York: Cambridge University Press, 1981.

Chalmers, Thomas Wightman. "Conduction of Electricity Through Gases." In *Historic Researches: Chapters in the History of Physical and Chemical Discovery*, edited by T. W. Chalmers, ch. 10. London: Morgan Brothers, 1949.

Chown, Marcus. "Just Who did Discover the Electron?" *New Scientist* (March 29, 1997): 49.

Crookes, William. "On the Illumination of Lines of Electrical Pressure, and the Trajectory of Molecules." Bakerian Lecture. *Philosophical Transactions of the Royal Society* 170 (1879): 135–64.

Dahl, Per F. *Flash of the Cathode Rays: A History of J. J. Thomson's Electron.* Bristol and Philadelphia: Institute of Physics Publishing, 1997.

DeKosky, Robert K. "William Crookes and the Quest for Absolute Vacuum in the 1870s." *Annals of Science* 40 (1983): 1–18.

Etter, L. E. "Some Historical Data Relating to the Discovery of the Roentgen Rays." *American Journal of Roentgenology and Radium Therapy* 56 (1946): 229.

Faraday, Michael. *Experimental Researches on Electricity*, vol. 1. London: R. and J. E. Taylor, 1839.

Forman, Paul. "The Discovery of the Diffraction of X Rays by Crystals: A Critique of the Myths." *Archive for History of Exact Sciences* 6 (1969): 38–71.

Fournier D'Albe, Edmund Edward. *The Life of Sir William Crookes, O.M., F.R.S.* London: T. F. Unwin, 1923.

Golinksi, Jan. *Making Natural Knowledge: Constructivism and the History of Science.* Cambridge and New York: Cambridge University Press, 1998.

Hittorf, Johann Wilhelm. "Ueber die Elektricitätsleitung der Gase." *Annalen der Physik und Chemie* 136 (1869). Translated in William Francis Magie, *A Source Book in Physics.* New York and London: 1935, 563.

Hughes, Thomas P. *Networks of Power: Electrification in Western Society. 1880–1930.* Baltimore: Johns Hopkins University Press, 1993.

Kragh, Helge. "J. J. Thomson, the Electron, and Atomic Structure." *Physics Teacher* 135 (1997): 328–32.

Molan, Charles. "Who Discovered the Electron?" *Technology Ireland* (October 1992): 56–59.

Morton, Alan Q. "The Electron Discovered 1897—J. J. Thomson's Apparatus." *Bulletin of the Scientific Instrument Society* 63 (1999): 29–30.

Morton, Alan Q., and Jane Wess. *Public and Private Science: The King George III Collection.* Oxford: Oxford University Press, 1993.

Müller, Falk. *Gasentladungsforschung im 19. Jahrhundert.* Berlin, 2004.

Noakes, Richard. "Telegraphy is an Occult Art: Cromwell Fleetwood Varley and the Diffusion of Electricity to the Other World." *British Journal for the History of Science* 32 (1999): 421–59.

Nollet, Abbé. *Leçons de physique expérimentale*. 6 vols. Paris, 1743–48.

Plücker, Julius. "On the Action of the Magnet upon the Electrical Discharge in Rarefied Gases." Translated by F. Guthrie. *The London, Edinburgh and Dublin Philosophical Magazine* 16 (1858): 119–35.

Robotti, Nadia. "J. J. Thomson at the Cavendish Laboratory: The History of Electric Charge Measurement." *Annals of Science* 52 (1995): 265–84.

Waller, John. *Fabulous Science: Fact and Fiction in the History of Scientific Discovery.* Oxford: Oxford University Press, 2002.

Weinberg, Steven. "The First Elementary Particle." *Nature* 386 (March 1997): 213–15.

Looking into (the) Matter

SCIENTIFIC ARTIFACTS AND

ATOMISTIC ICONOGRAPHY

Arne Schirrmacher

I T IS PROBABLY a myth that the history of science and the history of scientific objects are converging enterprises that eventually will coincide in one comprehensive historical account of the scientific endeavor. Clearly, the history of looking into matter of various kinds can be presented as a history of artifacts that allowed for new insights into these kinds of matter whether it was with a microscope, a NMR spectrometer, or a particle accelerator. Despite the fact this is what science museums could do best (and probably should), the prevailing mode of discourse subordinates the history of looking into matter to those of thinking about matter and of conceptualizing matter in general. In this way philosophy and imagination have often ruled over matters of fact.

To make things even more complicated recent historical scholarship has argued that alongside the perspectives of matter theories and their conceptual development on the one hand and that of experimentation involving scientific artifacts on the other hand, there is also a third point of view: the history of images or rather of atomistic iconography.[1] What separates these perspectives are the respective claims of autonomy, that is, that each one corresponds to an independent tradition not affected by more or less radical changes in one of the other fields. Theory tradition, iconographic tradition, and object tradition, for short, form different layers of scientific development with certain stabilizing connections much like the brick-wall metaphor of Peter Galison's history of particle physics.[2]

I will focus in this chapter on the question how the understanding of the nature of matter developed—mostly in the twentieth century—using the terms "looking-perceiving" and "image" in a wider sense. As for atoms they have been pictured as balls, modeled along the analogy of planetary systems of mechanical machinery or mentally perceived as ethereal structures or fields.

After a short sketch of the iconographic perspective and its claim of exhibiting invariant structures of knowledge I will ask to which extent also the history of artifacts could claim autonomy and the power to define knowledge structures for our understanding of matter. Before I can begin to deal with this question, however, I have to point to one more distinction regarding the status of artifacts. Two sorts must be separated that both inhabit our museums: those artifacts primarily manufactured for scientific research and those built as didactic means for mostly representing otherwise gained knowledge.[3]

Images (of) Matter

It is one of the main tasks of a historian of scientific objects to identify their contemporary roles within the scientific development and to remove the retrospective interpretations and categorizations applied to them as much as possible. This, however, also applies to the second type of artifacts. Take the "atomic models" manufactured for the department of atomic physics at the Deutsches Museum that demonstrate the atomic conceptions of Democritus and Lucretius. Do they reflect the atomic iconography of antiquity or the Roman Empire?

FIGURE 8.1
Atomic models made for display at the Deutsches Museum. Courtesy Deutsches Museum, Archive (DMA), BN R949/05 and 02.

Beyond doubt Democritus was one of the early atomists, though we do not have many genuine sources of his teachings. It is in the writings of his junior by three centuries, Lucretius, where we find the text that gave rise for the atomic models of figure 8.1. In the second chapter of his *On the Nature of Things* (De rerum natura) we read:

> Thus simple 'tis to see that whatsoever
> Can touch the senses pleasingly are made
> Of smooth and rounded elements, whilst those
> Which seem the bitter and the sharp, are held
> Entwined by elements more crook'd, and so
> Are wont to tear their ways into our senses,
> And rend our body as they enter in.[4]

Though this seems to be a clear account of atomic modeling, and the widespread German translations are even more unmistakably speaking of "smooth and rounded atoms," that is, it spells out the indivisibility property, a look at Lucretius's original lines shows immediately that the situation is at least not this clear. The term "element" appears in his educational poem, referring to elements like fire and water but also to some primordial objects (primordial rerum), but not at this point. There is even made mention of the possibility of "larger elements" in olive oil some lines before, thus contrasting with the primordial objects that for itself should be free of color, taste, and so on. Here the lines read:

> ut facile agnoscas e levibus atque rutundis
> esse ea quae sensus iucunde tangere possunt,
> at contra quae amara atque aspera cumque videntur,
> haec magis hamatis inter se nexa teneri
> proptereaque solere vias rescindere nostris
> sensibus introituque suo perrumpere corpus.[5]

Ea quae meaning *the ones that* (i.e., objects, items, things, maybe bodies, shapes, etc.), however, leaves much room for interpretation and obviously what we read in the English or German translations is to some extent not from Lucretius but from the translators' understanding of the atomism of his time.

The point I want to make here is not to claim that the model presented to the visitors of the Deutsches Museum is actually wrong, nor is it necessary to evaluate how much we can hope to learn from an educational poem that still ranks poetic form and language higher than a rather austere exactitude. It is rather that there were no models or images in the writings of Democritus or Lucretius! We have to understand (and probably also to "exhibit") that there were times in history without pictorial representations of the contemporary matter concepts.[6]

The Invention of Atomist Iconography

Only in recent years Christoph Lüthy has demonstrated convincingly that there was no atomistic iconography before the late sixteenth century. In particular his search for atomic representations in more than seventy editions of Lucretius text that were printed between the late fifteenth century and the early eighteenth century brought to light only a number of dramatic illustrations but no graphic representations of constituents of matter.[7]

What made the emergence of the globular atom impossible through the Aristotelian and scholastic tradition was a particular kind of anti-atomism. For Aristotle natural bodies appeared continuous and homogeneous and while they corresponded to certain "forms" this did not entail that pictures could be drawn, since these "forms"—as distinct from "figures" (figura)—meant logical principles, that is, forms of thinking, rather than graphical figures or images. Though geometry played some role, here again the idealized relations of the (mathematical) geometry of forms were of interest, not the physical geometry of nature. Scholastic tradition did not add many illustrations, which rather remained "notoriously few" but only "seemingly endless commentaries."[8] In the cases where images appeared, they graphically illustrated relations, inclusions, or hierarchical orderings, like the widespread onion-ring model of ordered inclusions.

With the Renaissance the Platonic view resurfaced that saw a correspondence of regular solids and the elements. While it is well-known that Kepler took up this structure to describe the proportions of the planetary orbits in the solar system, no convincing relation was created between shapes and substances. Plato's attempt to relate wedge-shaped pyramids with fire could not convincingly be extended by early modern thinkers.

Lüthy finally finds the full set of images of piled up globular atoms, the reference to Democritus, and the use of the term "atom" in Giordano Bruno's 1591 *De triplici minimo et mensura*. This birth of atomic iconography, however, did not coincide with a revolution of a related theory of matter. Rather one does find Bruno's new imagery enmeshed in old theological, arithmetical, and numerological speculations. Later natural philosophers like Kepler, Jungius, or Descartes took away much of this historical baggage and reinterpreted the new iconography within their theories and philosophies. Cutting short a complex story, Lüthy proposed the following thesis:

> The globular atom is an invention of the late sixteenth century. Neither did it exist before, nor did its invention seem very useful at first. Instead, the globular particle of matter is a strange outgrowth of Renaissance speculation which required decades of reinterpretation before it began to seem useful here and there as a possible tool for the explanation of certain natural phenomena.[9]

I will in the following try to show that a thesis of this kind can also be put forward for twentieth-century atomic imagery, when a similar invention of a new iconography has taken place, however, without being able to replace the globular atomic iconography fully in the public and in education, where it still lives on in its fifth century of existence.

Artifacts and New Images of Matter

Within the sciences new images of matter were clearly inevitable when experiment approached the realm of the atom. In particular the colorful phenomena of electrical discharges in evacuated glass tubes gave rise to a variety of images on what William Crookes called "the fourth state of matter." Interestingly, it was subatomic particles rather than atoms that became first visibly accessible through scientific artefacts.[10]

To illustrate the iconographic quality and persistence of the new images that originated in the late nineteenth century compare for a moment two typical images of matter from the beginning and from the end of the twentieth century, which were both available for any interested audience.

In the sixth edition of the popular German encyclopedia *Meyers Großes Konversationslexikon* that was published between 1906 and 1909, a full one-page colored plate with a number of drawings illustrated the entry on electrical discharge phenomena. The drawing

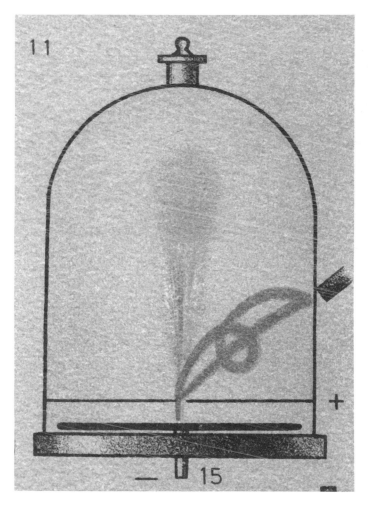

FIGURE 8.2
Drawing of discharge tube phenomena. Detail from a plate of the sixth edition of *Meyers Großes Konversationslexikon*, vol. V, 1906.

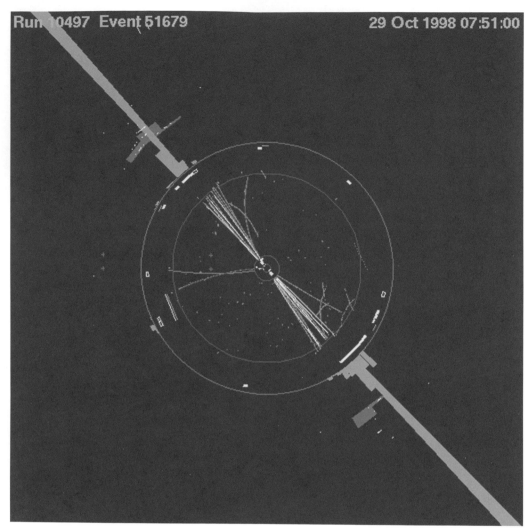

FIGURE 8.3
OPAL event 51679 of run 10497 from 1998. Courtesy CERN.

given in figure 8.2 is particularly telling as it foreshadows much of the typical iconographic elements that physicists were still employing a century later in computer-generated visualizations of particle accelerator experiments. In the *Konversationslexikon* one could read about this drawing:

> When one attaches closely above a disk-shaped cathode an also disk-shaped anode with a hole in the center, then from this aperture a pencil of rays comes out that decomposes into three parts when a magnet is approached: an uninfluenced pencil of canal rays and two pencils of cathode rays, one traveling in the directions of the lines of force, the other on a trajectory of a spiral around the former.[11]

Here we have, roughly speaking, all the elements of modern particle accelerator imagery: tracks coinciding in one point at different angles, their shape according to electric and magnetic fields and different colors identify different physical entities. The OPAL event shows a similar interaction of three lines, straight and spiral ones, that appear within a border given by the glass tube and the detector wall, respectively (see figure 8.3).

Hence we see that despite the great progress physics has made in the fields of atomic and particle physics during the twentieth century and despite the magnificent development in scale and power of the experimental machinery during this time we find again a surprisingly stable iconography at least for this kind of representations of matter. In order to make the relation between scientific artifacts and atomic iconography more explicit I will consider a number of typical experiments of atomic physics from the first quarter of the twentieth century, each of them involving a central scientific artifact.

I will place these objects into two groups, which I label as "looking at" and "looking into" approaches. Looking at nature or pieces of matter means inspecting and viewing. This mode corresponds to the shiny part of experimentation like the light phenomena of electric discharges. The representations are photographic, depicting and taking generally the phenomenon as a whole. Looking into matter as such or matter as a scientific problem, however, rather means exploring, investigating, studying, analyzing, preparing, or even constructing. Its mode is more representational, graphic, and selective, like the computer-generated and deliberately colored displays of particle accelerator events. For a first characterization one may claim that the looking-at approach concerns questions of *visibility* while the looking-into approach concerns *visualibility*, or the feasible ways of *visualizing*.

Looking At: Creating New Limits of Visibility

The puzzle of the existence of atoms and the microscopic (!) structure of matter has hardly benefited from the introduction and refinement of microscopes that revolutionized other fields, first of all biology. Only in the twentieth century, avenues were found to cross the borders of microscopic resolution that fell short of the atomic scale. I will consider two artifacts in order to discuss the question to what extent it was actually possible to shift the limits of visibility by inventing new ways of looking at matter and to what extent scientists hoped to be able to see even into the atom. Both originated in the first two decades of the twentieth century, hence well before the advent of the electron microscope in the 1930s, that today dominates the imagery of the atomic scale and that has in recent time received much historical interest, in particular regarding the questions whether its images are mere constructions from abstract data rather than representations and who is in control of the pictures.[12]

Seeing the Invisible: The Ultramicroscope

It is a telling coincidence that in the same year, 1872, when Ernst Abbe finished his theoretical work on image formation in microscopes Emil Du Bois-Reymond delivered his widely circulated *Ignorabimus* address at the Leipzig *Naturforscherversammlung*. While Abbe had arrived at firm foundations for his formula on the resolution limit of microscopes, which entailed that structures finer than a fifth of a micrometer could not be seen though any such optical device, Du Bois-Reymond contemplated on the limits of science in general, claiming finally that there are areas of knowledge besides the grasp of experimental research with scientific instruments: Not only were the mysteries of the human body and mind out of reach for the scientist, but also "confronted with the mysteries what matter and force were and how one could conceptualize them, he must once and for all settle upon the much harder acknowledgeable truth: 'Ignorabimus,'" that is, we will never know.[13]

It is, however, an equally telling coincidence that shortly after the mathematician David Hilbert strongly rejected the ignorabimus mentality in his 1900 Paris address and made this a constant theme of his public lectures, Abbe's firm tried to create a new type of microscope that would transgress the resolution limit.[14] The 1903 ultramicroscope (figure 8.4) of colloid chemist Richard Zsigmondy and Zeiss instrument maker Henry Siedentopf represents a successful combination of interests: While Zsigmondy needed instruments for specific studies of colloids, in particular in order to determine the size of the colloidal particles and to see whether kinetic theory would be applicable, the young physicist Siedentopf, one of quite a number of young university graduates mostly in physics hired by Zeiss around 1900, represented a new scientific culture in the Zeiss Werke that became closely related to scientific research questions ranging from colloids, which bore potential application in optics, to possibly the existence of atoms in general.[15]

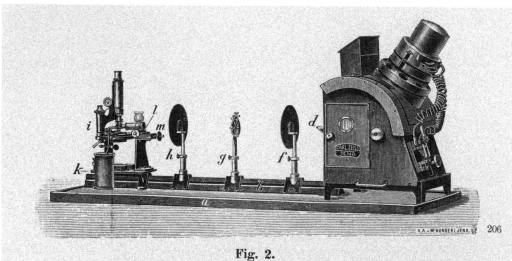

Fig. 2.

Einrichtung zur Beobachtung ultramikroskopischer Teilchen in Flüssigkeiten nach SIEDENTOPF und ZSIGMONDY.

FIGURE 8.4

1903 ultramicroscope. Ultramikroskopie für Kolloide. Nach Siedentopf und Zsigmondy (Zeiss Druckschrift Mikro 229), Jena, 3. ed. 1910, p. 5. Courtesy Deutsches Museum. The parts are labeled in the text as follows: (a) table, (b) optical bench, (c) projection arc-lamp, (d) aperture, (f) first projection lens, (g) precision slit head, (h) second projection lens, (i) microscope tripod, (k) ground plate, (l) cross sledge, (m) screw.

As it was the Göttingen colloid chemist who had realized that it should be possible to observe with a microscope perpendicular to the direction of illumination and against a dark background diffraction cones of particles smaller than Abbe's limit (a phenomenon known as Tyndall's effect), it took actually one and a half years and the full support of Zeiss optical know-how orchestrated by Siedentopf to realize the rendering visible (*Sichtbarmachung*) of colloidal particles not hitherto visible.[16] In principle the design of an ultramicroscope was quite simple, combining an ordinary microscope with an appropriate light source and a sample cell of favorable dimensions. Avoiding light to scatter into the dark background and focusing, however, turned out to be severe obstacles to be overcome by mechanical knowledge and skill; dark field condensers and the variant of the immersion ultramicroscope were developed in the following years.[17]

In their joint seminal paper Zsigmondy and Siedentopf reported that they "were able to make visible individual gold particles whose sizes were not very far from molecular dimensions."[18] Moreover, they demonstrated that—although strictly speaking what they observed were only diffraction disks, or cross-sections of widening cones of diffracted light—they could still make visible the individual particles, as they could be separated, traced, and also determined in size. The dimensions of the diffraction disks photographed, however, do not permit any conclusion about the actual particle size, which was calculated from relating counted particle numbers per area with specific weight and colloid concentration.[19]

Naturally, the two authors were most interested to communicate the application of their method and discussed at length how the color of gold ruby glasses depended on the size of the now rendered visible gold particles. This may explain why they mentioned that their ultramicroscope would be "especially appropriate" for the study of Brownian motion only in passing and why they omitted any statement about the feasibility to make visible single atoms and thus proving their reality.[20]

It is generally understood that it was the second of Einstein's three 1905 papers that raised this question forcefully and it was in particular Jean Perrin who convinced both science and the public of the reality of atoms.[21] Charlotte Bigg has stressed that the ultramicroscope as a symbolic artifact on the one hand and as a practical device that allowed the demonstration of the movement of ultramicroscopic particles in agreement with the kinetic theory on the other hand served Perrin to present visual evidence for the existence of atoms without actually showing images of them.[22] Neither were ultramicroscopic particles single atoms nor did the photographs or projections presented depict particles as particles.

Still, the ultramicroscope not only pushed the limit of visibility—though not far enough to make visible single atom—its nicely observable diffraction spots of single particles furthermore changed the standards of acceptability so that its audience more or less believed to have seen atoms. To demonstrate how this happened, one could analyze in some more detail the notions used for photographs of indirect imaging methods like ultramicroscopy, X-ray diffraction, or cloud chamber methods. In the popular German scientific monthly *Kosmos*, for example, a nice symmetrical Laue spot picture was presented in 1913 as "Atomphotogramm," thus suggesting that it would be something very much like a photographic image of a single

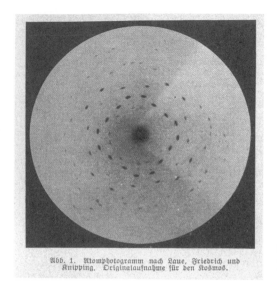

Abb. 1. Atomphotogramm nach Laue, Friedrich und Knipping. Originalaufnahme für den Kosmos.

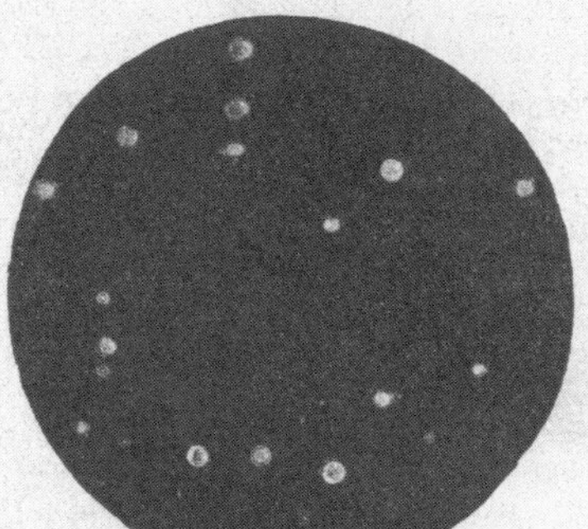

Abb. 5. Photographie ultramikroskopischer Goldteilchen einer Goldlösung. Durchschnittsgröße der einzelnen Teile 15 Millionstel Millimeter. (Aus Pöschl, Kolloidchemie.)

FIGURE 8.5
Images of Laue spots and the appearance of gold particles through an ultramicroscope. Kosmos 10 (1913), 265, and 14 (1917), 94.

atom; in 1917 a "photography of ultramicroscopic gold particles" of fifteen nanometers size again pretended to be a true photographic depiction of the particles (figure 8.5).[23]

There is no need to discuss the well-known Laue experiment here in detail, which basically turned to invisible radiation of smaller wavelength in order to shift the limit of resolution (and replaced the eye completely with the photographic film). It may suffice to recall that the 1912 experiment was generally taken as having demonstrated the atomic nature of crystals unequivocally, though the direct relation of the spot patterns in Laue photographs with the arrangement of the atoms in a crystal was anything but immediate and Laue and collaborators needed some time and collegial advice to arrive at the proper interpretation of their photographs.[24] What they saw was, mathematically speaking, the reciprocal lattice of the crystal lattice. So one might say that as in the 1917 *Kosmos* where the notion of "ultramicroscopial seeing" (ultramikroskopische Sehen) was coined, the Laue experiment introduced a kind of "reciprocal seeing," which still represents spatial relations but, for instance, falsifies symmetry patterns.

While the ultramicroscope like the Laue method meant an extension of the traditional observation concepts one had to put up with certain distortions or, to put it in a more positive way, one had to learn new ways of looking at nature. To which extent physicists believed to be able to push the limit of visibility will demonstrate my next example.

Hopes to Picture the Conceived: The Origins of the Debye-Scherrer Camera

The second artifact I would like to consider for the "looking at" category is the Deybe-Scherrer camera (figure 8.6), built in 1915 or 1916 to photograph—in some appropriate sense—the electron rings within the atoms, which Niels Bohr had proposed. In a paper presented to the

Göttingen Academy of Science in early 1915, Debye attempted nothing other than the "ultra-microscopy of the interior of the atom." Presenting a theoretical discussion of X-ray scattering at randomly orientated Bohr type atoms and combining it with an appropriate interpretation of the Laue experiment he convinced himself that "it must be possible in this way to establish by experiment the particular arrangement of the electrons in the atoms." And he concluded, "Whether, experimentally, rings are actually photographed or a continuous deviation from the scattering laws for dipoles is established" does not matter this much as long as "it appears to be essential that . . . we are in a position to measure from observations of the scattered radiation, the electron arrangement inside the atoms in centimeters."[25]

After this proposal, which may remind us strongly of the language and procedures for the ultramicroscope, what followed was a story of failure to photograph the electron rings or the positions within the atom and a reinterpretation of this failure into an innovative and successful method for X-ray structure analysis of specimens that do not allow for larger crystals, which would have made them accessible with the Laue method.

Again the apparatus was in principle quite simple: an X-ray tube and a camera containing X-ray film. As Paul Scherrer later recalled, however, a number of obstacles had to be overcome:

> Debye proposed to me that we try together such diffraction experiments. We used at first a gas-filled medical X-ray tube with platinum target which happened to be available in the collection of the institute. For power source we used an enormous induction coil with mercury interrupter and a gas-filled rectifying valve. The whole set-up appears nowadays like a show piece taken from a museum. The first diffraction photographs, with paper and charcoal as the scattering substances, showed no diffraction effects. The reason for this may have been that the thick glass wall of the tube absorbed the Pt L-radiation and transmitted only the continuous background. The film was relatively insensitive for the K-radiation, which, besides, was not strongly excited, so that possible maxima were covered up by the continuous background. This prompted me to construct a metal X-ray tube, water-cooled and with copper target. The tube remained connected to the rotating Gaede mercury pump. An aluminum window, 1/20 mm thick, permitted the rays to emerge. I also constructed a cylindrical diffraction camera, of 57 mm diameter, with a centering head for the sample, of the type which is being used still nowadays.[26]

The camera is basically a cylinder with X-ray sensitive film affixed to the wall all the way around. The sample has to be placed in the center and through a tube the monochromatic X-rays reach the sample. By interference the diffracted radiation shapes into cones that yield a pattern of curved lines on the film.[27]

Did this artifact allow photographing the electron rings in the atom? Debye and Scherrer reported one year later that:

> Experiments since then carried out by us show the expected result. However, in several instances, interference patterns of a different nature, and superimposed on the expected effect, were established, which indicated definitely by the sharpness of their maxima that the

FIGURE 8.6
Debye-Scherrer camera. Formerly exhibited at the Deutsches Museum as "Camera for photographing powder diagrams with X-rays. Original from P. Debye and P. Scherrer, 1917." It was provided to the Deutsches Museum in January 1920 by H. Debye. Courtesy Deutsches Museum.

regular arrangement of the presumably small number of electrons in the atom cannot be held responsible for their occurrence. The present preliminary publication will be restricted to the description and explanation of this phenomenon. In a later publication we intend to treat the electron interferences.[28]

In quite a number of subsequent articles, which I cannot analyze here in detail, the authors more or less diffused their promise to come back to the "ultra-microscopy of the interior of the atom" and promoted their "method for the determination of the atom arrangement in crystals" instead.[29] When the *Handbuch der Experimentalphysik* covered in its 1928 volume on structure analysis by X-ray interferences also the work of Debye and Scherrer, reference was given only to a specific later publication in which the original aim of the experiments had disappeared.[30]

Debye knew what he wanted to see when looking at matter with his experiment. The photographic film, however, could not make visible what it should. Comparing this case to the later imagery electron microscopes provided, it may seem that the complex and widely adjustable ways in which these new devices produced images included one mode of representation that met the liking of the experimenter: landscapes of single atoms.[31] This brings us to our second category of the looking-into approach, since the electron microscope is already a hybrid rather than a pure looking-at device.

When Ian Hacking asked, "Do we see through a microscope?" he basically concluded, "Yes." Acknowledging differences between microscopic and macroscopic seeing,[32] for example, by the effects of diffraction, he still suggested that realism and the independent interference with various methods of otherwise not visible structures provide good reason to believe in the expansion of visibility. There may be some unclear territories like the question whether we can accept that the habit of crystallographers to discuss all their physics in reciprocal space (to which, e.g., the Laue photographs relate) amounts to seeing their specimens in such an alternative space.[33] Nonetheless the criteria mentioned for the looking-at category are met, which comprise photographic nature, depicting quality and covering the entity as a whole.

Looking Into: New Attempts of Visualization

Using a distinction Giora Hon introduced, the second type of experiments for exploring matter is related to the "bombardment method." According to Hon, around the beginning of the twentieth century experimental physics underwent a "transition from the study of propagation phenomena to questions of structure" that was "reflected directly in the development of a new experimental technique that was conceived when physics turned its attention from macro- to microphysical problems." This bombardment method emerged "when it became clear that rays and particles of known properties could be manipulated and used as probes that could impinge on, collide with, or plunge through the object under study."[34] How much this bombardment method changed the understanding of the atom early in the atomic century can be seen from my third artifact.

From Absorption Measurements to the Empty Atom: Lenard's Cathode Ray Tube

Claiming that a modern particle accelerator is in principle nothing else but the old cathode ray tube of the nineteenth century, one may ask since when did such a bold equation make sense and since when did scientists understand their tubes as particle stream sources that can be used to probe matter. First of all, cathode rays were imprisoned in their glass tubes and, corresponding to the nature of the (necessary) gas filling of the tube, a wide variety of colorful, but hard to describe and to classify, phenomena occurred.[35] It was Philipp Lenard's merit to free the rays from their tube and let them penetrate a thin aluminum "window" and thus be available as pure rays that could serve for many purposes.[36] How the rays escaped from the 1894 Lenard tube was far from clear. Was it oscillations like sound waves that could go through a membrane, was it phenomena of the immaterial, all-penetrating ether, or was it corpuscles or electrons, like the British physicists liked to believe?

Lenard had combined his cathode ray tube with an observation tube so he could study the properties of the rays in vacuum, electric, and magnetic fields and in the presence of arbitrary substances (figure 8.7). When he started in 1895 to study systematically the absorption of cathode ray by many kinds of matter, ranging from hydrogen gas, paper, and glass to mica, aluminum, and gold, he was not yet convinced of the particle nature of his rays. Thus his universal law about the absorption of cathode rays, which he found, did not immediately relate to atomic theory of matter. The empirical law stated that the absorption power

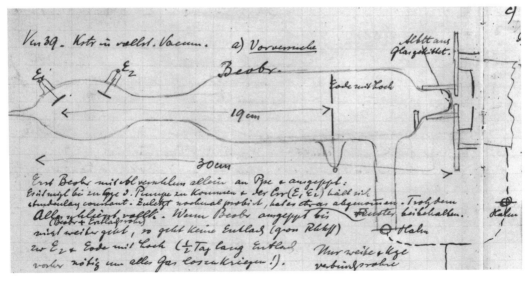

FIGURE 8.7
Lenard tube of 1894 (top) and drawing from laboratory notebook dated December 22, 1892 (bottom). Courtesy Deutsches Museum.

of any substance is independent of its particular physical or chemical properties and only depends on its density.

But when in the following years the electron emerged as a reality with measurable mass, velocity, and charge, the propagation of rays became bombardments with particles. Lenard's measurements, which he extended in the following ten years, now meant through the particle interpretation that (1) the electrons of the cathode rays can travel through thousands of atoms without absorption, that (2) the rate of absorption depends merely on the density of matter, and that (3) only for very slow electrons a higher than predicted absorption takes place, which points at electric forces in the atoms. In this way Lenard concluded from clear and undisputable experimental findings that atoms are almost completely empty, that their stability had to do with electric forces, and that atoms could possibly consist of one type of primary matter that he called *dynamids*. In his seminal 1903 paper he stated:

> For example, the volume in which one finds one cubic meter of solid platinum is empty— in the same sense like the cosmic space, that is traversed by light—save for *at most a cubic millimeter* as the complete true dynamide [corpuscle] volume.[37]

Lenard's "empty" atom had most of the parts Rutherford's was later celebrated for.[38] The only thing Lenard could not see with his electron bombardment method was whether the positive charge was concentrated in the center of the atom or whether pairs of positive and negative charge would fill the atoms, the alternative for which he opted.[39]

Lenard was probably the first to put forward the new paradigm for looking into matter, when he told the audience of his Nobel speech, which was later published in two editions, that "we can employ the quanta of the cathode rays as tiny probes, which we let pass through the interior of the atoms, so that they provide us with knowledge of this interior."[40] In this way, there is good reason to take Philipp Lenard, at the turn of the twentieth century, as the founding father of this tradition rather than Rutherford ten years later, the more so as Lenard also introduced specific notions that originated from his absorptions researches like "cross-section" for describing the probability to scatter particles at a target.

Despite accurate numbers about impenetrable volumes and absorption behaviors, the new insights into the atom did not give rise to a clear picture. Like the old atomists, the knowledge of the atom—now experimental rather than philosophical—was nonpictorial:

> We are amazed at seeing, that we have got beyond the old impenetrability of matter. Every atom of matter claims in fact an impenetrable space with regard to the others; but with respect to the free quanta of electricity *all sorts of atoms* prove to be *pervious structures, like built up from finer constituents with much space in between*.[41]

This does not mean that no models for the atom were discussed. Already around the time of the news of Lenard's findings the planetary analogy was cited by various authors in more popular journals—even by Lenard via his assistant—but for physicists it was clear that the mechanical equilibrium that held for gravitation did not exist for the electrical forces that would immediately slow down a circulating electron. As a consequence, no drawings of these atomic conceptions appeared and it might be worthwhile to mention in this context that also Thomson's atomic model, now so prominent under the title of plum-pudding model, was neither put forward seriously with illustrations nor was it given this title at the time.[42] What happened to establish the planetary atom was a rather long negotiation process between science and public that eventually came to agree on accepting Bohr's quantum physical extension of the mechanical analogy, a development I will return to in some more detail in the conclusion.[43] Before I do, I would like to turn to my last artifact that shows how the experimental knowledge of the empty atom, even after Rutherford, was taken to contradict Bohr's atom rather than to support it.

From Refuting to Substantiating the Bohr Atom: The Franck-Hertz Experiment

As mentioned before, neither the Rutherford atomic model nor Bohr's was an immediate success. Looking through the leading German abstracting journal that aimed at communicating scientific news between the specialists of different science fields, *Naturwissenschaften*, for the years between 1913 and 1916 one can find quite a variety of ideas about the atom, however, without any preference to Bohr or Rutherford.[44] Hence, it is no surprise that James Franck and Gustav Hertz started around this time their experimental researches on the atom with Lenard's tubes and methods from his 1902 paper on the photoelectric effect, the very same paper Einstein cited as the experimental basis of his 1905 light quanta paper,[45] and then pursued them with the general aim to check "the relations which emerge both from the quantum theory and the considerations of atomic models."[46] Similar to the Debye-Scherrer case, Franck and Hertz knew what they wanted to see: ionization of molecules by bombardment with electrons carrying a certain amount of energy, the so-called ionization energy. For this purpose electrons were accelerated by an electric field within a tube filled by low-pressure mercury vapor (Fig. 8.8).

Summarizing their findings, Franck and Hertz wrote in May 1914 that an energy transfer to the mercury molecules by 4.9 volt electrons resulted in their ionization.[47] Electrons of less energy showed elastic scattering, those with double the threshold voltage were able to ionize two mercury molecules, and so on. Implicitly, the authors presupposed that mercury molecules contain electrons but they did neither assume them to form a specific structure in the atoms nor to have levels of binding to the molecule other than that type that can be destroyed by the ionization process. Like Debye and Scherrer who did not see the electron rings, Franck and Hertz did not see Bohr's energy levels of the atoms.

Bohr, who immediately recognized the support the experiment would lend to his theory, explained in a 1915 paper that the correct interpretation of the experiment would be to understand the energy threshold as that of the transition between ground state and first exited state of an electron within the mercury atom rather than of ionization that should take place only for much higher voltage. Franck and Hertz, however, did not correct their interpretation but moreover rejected Bohr's view and challenged his whole theory instead.[48]

It took Franck and Hertz several more years to acknowledge that their apparatus did not produce any ionization. Meanwhile the new Bohr-Rutherford atom found more and more widespread acceptance and the first drawings of it appeared in journals after World War I and a wood and metal exhibit was made for the Deutsches Museum (Fig. 8.9).[49]

Apart from attributing a certain blindness to Franck and Hertz, a closer look at their measurements shows that, contrary to Bohr's predictions of a number of transitions between the many energy levels of the atom, solely the first one could be seen. Only much later did improved experiments show more of the structure of the atom. This reminds us of the main characteristic of the looking-into approach that is so clearly represented by the bombardment method: It is very selective of certain properties of matter and does in no way give an account of the whole object under investigation any more. This selectivity, however, also opens much room for choices of visualizations.

Conclusion: Visibility Lost and Visualization Regained?

In a now classic article of Arthur I. Miller on the genesis of quantum theory, which was expanded in a number of further publications through the last twenty-five

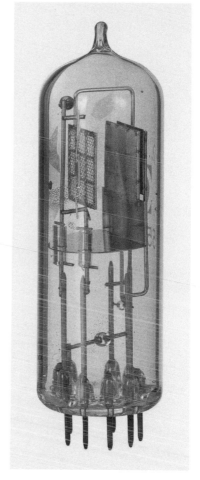

FIGURE 8.8
Franck-Hertz tube. Courtesy LD Didactic GmbH.

years, the thesis was put forward that around 1913 *visualization* was lost but it was regained around 1927.[50] As this account comports well with the traditional historiography of physical theory development—from Bohr's atom to the Copenhagen interpretation of quantum mechanics—it neither resonates convincingly with the experimental history of physics nor with the history of public communication about advances in this field.[51]

Shifting the point of view to these latter two directions I would like to propose a rather different thesis, which takes much from Lüthy's thesis presented in the introduction and from the two concurrent developments I discussed, the one of extending visibility and the one of creating new visualizations. With new discoveries on radiation and instability of matter in the last decade of the nineteenth century, the globular atomic iconography disappeared

FIGURE 8.9
Model of the hydrogen atom
made for the Deutsches Museum
from a concept of Arnold Sommerfeld and artistic advice by the
architect Friedrich von Thiersch.
Courtesy Deutsches Museum.

from the scientific discourse and was at least obscured in public recognition. While around
1900 the main aspects of the architecture of the modern atom became experimentally known,
no new picture of the atom was established until the end of World War I. Hence, we find
between 1895 and 1918 a period in the history of science devoid of any reasonable atomic
iconography. The new picture of the atom became more and more widely used in the following years mostly in the literature aimed at an interested public. In this way the empty
planetary atom—as a constructed visualization—became generally accepted just when quantum mechanics disproved the existence of electron orbits.

As it may have become already clear, attempts at explanation of this pictureless period
have to go beyond a history of scientific ideas or laboratory work. Also probably the first
suggestive model, the Sommerfeld model of 1918, was one created for and with the public.
The development may be viewed as a period of multiple superpositions of conflicting developments: Clearly, there was a superposition of attempts to shift the limits of visibility and
others to create new types of visualization replacing the visible. But it took place also in Ger-

many, where all four artifacts, which I presented, were employed and the related experimental researches were pursued, and Germany perhaps contributed most to create a new atomic physics. Here we also find a superposition of a well-structured Kaiserreich society with certain expectations with regard to science on the one hand and the emergence of a modern physics leaving behind most of the concepts of classical physics on the other hand, which had, as one may try to argue, a certain relation to or immersion in this particular culture. While, for example, physicists like Max Born and Alfred Landé realized during the 1918 revolution in Berlin that not only the political system would change within days, but also the Bohr-Sommerfeld model for the atom could not live on, since experimental results demonstrated the three-dimensional distribution of electrons in the atom contrary to the planar models, at the same time a disillusioned public was seemingly ready to allow for the lack of solidity and impenetrability of matter and to accept just this new picture presented in various articles in popular science journals through the 1920s.[52]

These remarks may suffice to show how complex this particular episode in the history of science actually becomes, when freed from the pure internalistic perspective. Since it is here not the place to tell this story more fully, I would like to conclude my paper with three points on physics, artifacts, and museums.

(1) As Giora Hon has argued convincingly, the progress quantum theory made in explaining the atom in the first two decades of the twentieth century were only possible when propagation experiments like the researches on black-body radiation and spectroscopy were combined with bombardment experiments like those of Lenard, Rutherford, and Franck and Hertz.[53] In this article I argued that this development is also mirrored in the disappearance and later in the establishing of a new atomic iconography. In writing a history of physics we should therefore not deny the existence of this transitory period of ambiguity and superposition of different, at times contradictory, experimental and theoretical approaches and findings. In this way also the projects of pushing the limits of visibility and of creating new ways of visualization were concurrent parts of this development, both necessary to give rise to modern atomic physics.

(2) The key to illuminating the process of establishing new theories, new models, and new images in science lies in the artifacts. Microscopes and discharge tubes bridge the fractures in interpretation, theory, and iconography. It is probably worthwhile to rank higher the "Atomphotogramme," Laue spot photographs, or ultramicroscopic pictures as compared to the often retrospectively constructed plum-pudding models, electron rings, or particle accelerator events on the computer screen. Examples like the Debye-Scherrer camera or the Lenard tube show that telling the stories of artifacts exhibits a great extent of autonomy that can be distinguished from those of theory and imagery.

(3) It should be a special challenge, and probably also a definite chance, for science museums to communicate also ambiguous scientific times like the pictureless periods of the atom. But can one build a physics exhibit about the emerging quantum theory (or ancient atomism) without the typical images? At least, I would argue, one should start telling the histories of the main experiments that led to modern atomic physics without squinting at school textbooks, the infallibility of scientific heroes, or the straightforwardness of scientific progress.

If there is a field where the slogan "visibility lost and visualization regained" is to be read as a warning, it is probably in the science museum: We should rather try to extend the visibility—in particular of artifacts—in the museum, rather than to content ourselves with finding selective and, at times, manipulating visualizations.

Notes

1. See, for example, Lefèvre, et al., *The Power of Images*; Miller, "Imagery and Representations"; and Latour and Weibel, *Iconoclash*.

2. Lefèvre, et al., write: "the striking independence of this tradition of visual representations from specific theories of matter . . . points to the fact that these theories comprise structures of knowledge invariant with respect to the great conceptual revolutions of science." See, Lefèvre, *The Power of Images*, viii; Galison, *Image and Logic*.

3. Clearly, this is not a sharp distinction; artifacts can also change their role between these two categories. The balls and wire construction kits widely employed in chemistry in the last third of the nineteenth and the twentieth centuries may serve as an example for an ambiguous object. The table croquet balls of August Hoffmann (in the Museum of the History of Science in Oxford) or the colored cardboard models of Jacobus van't Hoff (in the Deutsches Museum) clearly had a different status than the metal plates and rods of Watson and Crick in the double-helix structure (in the London Science Museum). See Meinel, "Molecules and Croquet Balls"; Sichau, "Atome. Eine lange Geschichte"; Chandarevian, "Portrait of a Discovery."

4. Lucretius, *On the Nature of Things*, in the often reprinted translation of William E. Leonard.

5. Ibid. Lines 402–7 of the second book.

6. This point clearly pertains to the recent work on the historicity of basic epistemological notions of science like fact, objectivity, and rationality. See, for example, the collection of articles by Daston, *Wunder, Beweise und Tatsachen*, and the current research focus on the history of observation at the Berlin Max Planck Institute for the History of Science.

7. Lüthy, "The Invention of Atomist Iconography," 122. Here one illustration from 1683 is discussed in some detail, which might qualify as depiction of atoms, but then is dismissed.

8. Murdoch, *Album of Science*, x; cited by Lüthy, "The Invention of Atomist Iconography," 13.

9. For Bruno's arguments and the reception of his writings see Lüthy, "The Invention of Atomist Iconography," 123f, quote on 118.

10. I focus here mainly on the point of view physicists took toward the atom. For the respective approaches and interests in rich history of chemistry see, for instance, Meinel, "Molecules and Croquet Balls."

11. *Meyers Großes Konversationslexikon*, quote on 614. The drawings probably originate from Otto Lehmann, cp. his *Die elektrischen Lichterscheinungen*.

12. For example, Rasmussen who, in *Picture Control*, demonstrates how involved the production even of seemingly plain pictures was. For this question with respect to the ordinary microscope see note 33.

13. On Abbé's research and publications, see Cahan, "The Zeiss Werke," 72f. Du Bois-Reymond, *Über die Grenzen des Naturerkennens*, 51: "Gegenüber den Räthseln der Körperwelt ist der Naturforscher längst gewöhnt, mit männlicher Entsagung sein 'Ignorabimus' auszusprechen. . . . Gegenüber dem Räthsel aber, was Materie und Kraft seien, und wie sie zu denken vermögen, muss er ein für alle mal zu dem viel schwerer abzugebenden Wahrspruch sich entschließen: 'Ignorabimus.'"

14. Hilbert, *Mathematische Probleme*. Consider also his 1930 radio address as discussed in Vinnikov, "We Shall Know: Hilbert's Apology." Cahan, "The Zeiss Werke," 86f.

15. Cahan, "The Zeiss Werke," 90f.

16. Ibid for details. For a brief sketch of the technical development in the following years see Ede, "Microscope, Ultra-."

17. Ibid., 400.

18. Zsigmondy and Siedentopf, "Über Sichtbarmachung," 2.

19. Cahan, "The Zeiss Werke," 94f.

20. Zsigmondy and Siedentopf, "Über Sichtbarmachung," 10.

21. Einstein, "Über die von der molekularkinetischen Theorie der Wärme." Perrin, "Agitation moleculaire," describes the projection of Brownian motion for public showing. In Perrin, "Mouvement brownien," he deals comprehensively with the reality question. See also Nye, *Molecular Reality*.

22. Bigg, "Brownian Motion."

23. Sieverking, "Sichtbarmachung der Moleküle," 268; and Kahn, "Das Ultramikroskop," 94.

24. Ewald, "Max von Laue," 137.

25. Debye, "Zerstreung von Röntgenstrahlen."

26. Scherrer, "Personal Reminiscences," 642f.

27. Von Miller to Debye.

28. Debye and Scherrer, "Interferenzen an regellos orientierten Teilchen I," 51f. In Scherrer's recollection things read differently: "Debye and I were most surprised to find on the very first photographs the sharp lines of a powder diagram, and it took us not long to interpret them correctly as crystalline diffraction on the randomly oriented microcrystals of the powder. The diffraction lines were much too sharp that they could have been due to the few scattering electrons in each single atom." Scherrer, "Personal Reminiscences," 643.

29. See Scherrer, "Das Raumgitter des Aluminiums," 23.

30. Ott, *Strukturbestimmung mit Röntgeninterferenzen*, 175, refers only to Debye and Scherrer, "Interferenzen an regellos orientierten Teilchen III." This paper starts with a renewed description of the method in which silently the term "ultramicroscopy of the interior of the *atom*" as used in the first publication, was replaced with "ultramicroscopy of the interior of the *molecule*," on 291.

31. For a discussion of this point for the case of the more modern scanning tunneling microscope see Hennig, "Versinnlichung des Unzugänglichen Oberflächendarstellungen."

32. The position that microscopical seeing was fundamentally different from macroscopical was already discussed intensely in the first half of the nineteenth century, as by Schickore, "Ever-Present Impediments."

33. Hacking, "Do we see through a microscope?" (See also the comments of Bas van Fraassen in Churchland and Hooker, *Images of Science*, 297–301.) The case of reciprocal space is discussed on 150. The invitation to the experimenter to verify what he or she sees by interference ("you learn to see through a microscope by doing, not by looking," on 136) is a central issue of Hacking's philosophy of science, as in Hacking, *Representing and Intervening*.

34. Hon, "From Propagation to Structure," 152. While Hon judges an early paper of Rutherford and J. J. Thomson in 1896 on the effect of X-rays on the conduction of electricity in gases as the beginning of the bombardment methods (p. 153f.), I would prefer to argue that this should rather be associated with matter particles. Clearly, it does not make sense to create a priority conflict between Rutherford and Lenard here as their agendas were far too different, but the suggested understanding of Rutherford's 1896 experiments as "bombardment" of electrons with X-ray particles seem to me inconclusive.

35. For example, Müller, *Gasentladungsforschung im 19. Jahrhundert*.

36. Lenard, "Über Kathodenstrahlen in Gasen."

37. Lenard, "Über die Absorption von Kathodenstrahlen," 739.

38. Although only after Bohr combined it with quantum theory; see, for the slow reception of Rutherford's paper, Heilbron, "The Scattering of *a* and *b* Particles," 300.

39. For a more detailed discussion, see Schirrmacher, "Das leere Atom."

40. Lenard, *Über Kathodenstrahlen*, 189.

41. Ibid.; a similar passage can be found some years earlier in Lenard, "Über die lichtelektrische Wirkung," 192.

42. Martinez, in "Plum Pudding and the Folklore of Physics," demonstrated that the first published account of "plum pudding" came nearly forty years later in a textbook and hence was related to a shift in the manners of physics teaching.

43. For a detailed account of this development, see Schirrmacher, "Der lange Weg."

44. For details see ibid.

45. In Franck and Hertz, "Über Zusammenstöße," they refer to the methods in Lenard, "Über die lichtelektrische Wirkung," as is done in Einstein, "Über einen die Erzeugung."

46. Franck and Hertz, "Über Zusammenstöße," 458.

47. Ibid., 466.

48. Bohr, "On the Quantum Theory of Radiation." Franck and Hertz, "Über Kinetik von Elektronen." For a brief account see Heilbron, "Lectures," 74–78; more detailed in Hon, "Franck and Hertz."

49. Schirrmacher, "Das leere Atom,"146f.

50. Miller, "Visualization Lost and Regained." Also, by the same author, *Imagery in Scientific Tthought*; and "Imagery and Representations."

51. Hon, "From Propagation to Structure," and Schirrmacher, "Der lange Weg."

52. Born to Hilbert; Schirrmacher, "Der lange Weg."

53. Hon, "From Propagation to Structure,"168f.

References

Bigg, Charlotte. "Evident Atoms: Visuality in Jean Perrin's Brownian Motion Research." *Studies in History and Philosophy of Science* 39 (2008): 312–32, 320.

Bohr, Niels. "On the Quantum Theory of Radiation and the Structure of the Atom." *Philosophical Magazine* 30 (1915): 394–415.

Born, Max, to David Hilbert. Letter, November 14, 1918. In Hilbert Papers, Staats- und Universitäts-bibliothek Göttingen, Box. 40A, Nr. 18.

Cahan, David. "The Zeiss Werke and the Ultramicroscope: The Creation of a Scientific Instrument in Context." In *Scientific Credability and Technical Standards in 19th and Early 20th Century Germany and Britain*, edited by Jed Z. Buchwald. Dordrecht: Kluwer Academic Publishers, 1996, 67–115.

Chandarevian, Soraya. "Portrait of a Discovery: Watson, Crick and the Double Helix." *Isis* 94 (2003): 90–105.

Churchland, Paul M., and Clifford A. Hooker, eds. *Images of Science, Essays on Realism and Empiricism, with a Reply from Bas C. van Fraassen*. Chicago: University of Chicago Press, 1985.

Daston, Lorraine. *Wunder, Beweise und Tatsachen. Zur Geschichte der Rationalität*. Frankfurt: Fischer Verlag, 2001.

Debye, Peter. "Zerstreung von Röntgenstrahlen." *Annalen der Physik* 46 (1915): 809–23. English translation in *The Collected Papers of Peter J. W. Debye*. New York: Interscience Publishers, 1954, 40–50.

Debye, Peter, and Paul Scherrer. "Interferenzen an regellos orientierten Teilchen im Röntgenlicht I." *Physikalische Zeitschrift* 17 (1916): 277–83. Translated in *The Collected Papers of Peter J. W. Debye*. New York: Interscience, 1954, 56–62.

Debye, Peter, and Paul Scherrer. "Interferenzen an regellos orientierten Teilchen im Röntgenlicht III (Über die Konstitution von Graphit und amorpher Kohle)." *Physikalische Zeitschrift* 18 (1917): 291–301.

du Bois-Reymond, Emil. *Über die Grenzen des Naturerkennens*. Leipzig: Veit, 1872 quoted from 7th ed. 1851.

Ede, Andrew. "Microscope, Ultra-." In *Instruments of Science. An Historical Encyclopedia*, edited by Robert and Deborah Warner. New York and London: Garland Publishing, 1998, 400–401.

Einstein, Albert. "Über die von der molekularkinetischen Theorie der Wärme geforderte Bewegung von in ruhenden Flüssigkeiten suspendierten Teilchen." *Annalen der Physik* 17 (1905): 549–60.

Einstein, Albert. "Über einen die Erzeugung und Verwandlung des Lichtes betreffenden heuristischen Gesichtspunkt." *Annalen der Physik* 17 (1905): 132–48.

Ewald, Paul Peter. "Max von Laue 1879–1960." *Biographical Memoirs of Fellows of the Royal Society* 6 (1960): 135–56.

Franck, James, and Gustav Hertz. "Über Zusammenstöße zwischen Elektronen und den Molekülen des Quecksilberdampfes und die Ionisierungsspannung desselben." *Deutschen Physikalischen Gesellschaft* 16 (1914): 457–67.

Franck, James, and Gustav Hertz. "Über Kinetik von Elektronen und Ionen in Gasen." *Physikalische Zeitschrift* 17 (1916): 409–16, 430–40.

Galison, Peter. *Image and Logic. A Material Culture of Microphysics*. Chicago: University of Chicago Press, 1997.

Hacking, Ian. *Representing and Intervening: Introductory Topics in the Philosophy of Science*. Cambridge: Cambridge University Press, 1983.

Hacking, Ian. "Do We See through a Microscope?" In *Images of Science. Essays on Realism, and Empiricism, with a Reply from Bas C. van Fraassen*, edited by Paul M. Churchland and Clifford A. Hooker. Chicago and London: University of Chicago Press, 1985.

Heilbron, John L. "The Scattering of α and β Particles and Rutherford's Atom." *Archive for History of Exact Sciences* 4 (1968): 247–307.

Heilbron, John L. "Lectures on the History of Atomic Physics 1900–1922." In *History of Twentieth Century Physics*, by Charles Weiner. New York: Academic Press, 1977, 40–108.

Hennig, Jochen. "Versinnlichung des Unzugänglichen. Oberflächendarstellungen in der zeitgenössischen Mikroskopie." In *Konstruierte Sichtbarkeiten. Wissenschafts- und Technikbilder seit der Frühen Neuzeit*, by Martina Heßler. München: Fink Verlag, 2006, 99–116.

Hilbert, David. *Mathematische Probleme, Archiv für Mathematik und Physik* 3. Reihe, Bd. 1 (1901). Reprinted in David Hilbert, *Gesammelte Abhandlungen*, vol. 3. Berlin: J. Springer, 1935, 290–329.

Hon, Giora. "Franck and Hertz versus Townsend: A Study of Two Types of Experimental Error." *Historical Studies in the Physical Sciences* 20 (1989): 79–106.

Hon, Giora. "From Propagation to Structure: The Experimental Technique of Bombardment as a Contributing Factor to the Emerging Quantum Physics." *Physics in Perspective* 5 (2003): 150–73.

Kahn, Fritz. "Das Ultramikroskop." *Kosmos* 14 (1917): 90–95.

Latour, Bruno, and Peter Weibel, eds. "Iconoclash. Beyond the Image Wars in Science, Religion and Art on the Occasion of the Exhibition Iconoclash." Karlsruhe, Germany: ZKM, Centre for Art and Media; Cambridge, MA: MIT Press, 2002.

Lefèvre, Wolfgang, Jürgen Renn, and Urs Schöpflin, eds. *The Power of Images in Early Modern Science*. Basel and Boston: Birkhäuser, 2003.

Lehmann, Otto. *Die elektrischen Lichterscheinungen oder Entladungen, bezeichnet als Glimmen, Büschel, Funken und Lichtbogen, in freier Luft und in Vacuumröhren.* Halle, 1898.

Lenard, Philipp. "Über Kathodenstrahlen in Gasen von atmosphärischem Druck und im äußersten Vacuum." *Annalen der Physik und Chemie* 51 (1894): 225–67.

Lenard, Philipp. "Über die lichtelektrische Wirkung." *Annalen der Physik* 8 (1902): 149–98.

Lenard, Philipp. "Über die Absorption von Kathodenstrahlen verschiedener Geschwindigkeit." *Annalen der Physik* 12 (1903): 714–44.

Lenard, Philipp. Über Kathodenstrahlen. Nobel-Vorlesung gehalten in öffentlicher Sitzung der Königl. Schwedischen Akademie der Wissenschaften zu Stockholm am 28. Mai 1906. Leipzig: J. A. Barth, 1906. Reprinted in *Philipp Lenard: Wissenschaftliche Abhandlungen aus den Jahre 1886–1932*, vol. 3, edited by Ludwig Wesch. Leipzig: S. Hirzel, 1944, 167–97.

Lucretius. *On the Nature of Things.* Translated by William E. Leonard. www.classics.mit.edu/Carus/nature_things.html (accessed 30 June 2009).

Lüthy, Christoph. "The Invention of Atomist Iconography." In *The Power of images in Early Modern Science*, edited by Wolfgang Lefèvre, Jürgen Renn, and Urs Schöpflin. Basel and Boston: Birkhäuser, 2003, 117–38.

Martinez, Ruben. "Plum Pudding and the Folklore of Physics." Presentation at History of Science Society Annual Meeting, Cambridge, MA, November 20–34, 2003.

Meinel, Christoph. "Molecules and Croquet Balls." In *Models: The Third Dimension of Science*, edited by Soraya de Chadarevian and Nick Hopwood. Stanford: Stanford University Press, 2004, 242–76.

Meyers Großes Konversationslexikon, vol. 5, 6th ed. Leipzig and Wien, 1906.

Miller, Arthur I. "Visualization Lost and Regained: The Genesis of the Quantum Theory 1913–1927." In *On Aesthetics in Science*, edited by J. Wechsler. Cambridge MA: MIT Press, 1978, 73–101.

Miller, Arthur I. *Imagery in Scientific Thought: Creating Twentieth Century Physics.* Boston: Birkäuser, 1984.

Miller, Arthur I. "Imagery and Representations in Twentieth-Century Physics." In *The Cambridge History of Science*, vol. 5, *The Physical and Mathematical Sciences*, edited by Mary Jo Nye. Cambridge: Cambridge University Press, 2003, 191–215.

Müller, Falk. *Gasentladungsforschung im 19. Jahrhundert.* Berlin and Diepholz: 2004.

Murdoch, John E. *Album of Science: Antiquity and the Middle Ages.* New York: Scribner, 1984.

Nye, Mary Jo. *Molecular Reality: A Perspective on the Scientific Work of Jean Perrin.* London: Macdonald; New York: American Elsevier, 1972.

Ott, Heinrich. *Strukturbestimmung mit Röntgeninterferenzen.* Vol. 7/2 of *Handbuch der Experimentalphysik*, edited by Wilhelm Wien. Leipzig: Verlags, 1928.

Perrin, Jean. "Agitation moleculaire et mouvement brownien." *Compte Rendus de l'Académie des Sciences* 146 (1908): 967.

Perrin, Jean. "Mouvement brownien et realité moléculaire." *Annales de Chimie et de Physique.* 8 ser. 18 (1909): 1–114.

Rasmussen, Nicolas. *Picture Control. The Electron Microscope and the Transformation of Biology in America, 1940–1960*. Stanford: Stanford University Press, 1997.

Scherrer, Paul. "Das Raumgitter des Aluminiums." *Physikalische Zeitschrift* 19 (1918): 23–27.

Scherrer, Paul. "Personal Reminiscences." In *Fifty Years of X-Ray Diffraction*, by Paul P. Ewald. Utrecht: Oosthoek, 1962, 642–46.

Schickore, Jutta. "Ever-Present Impediments: Exploring Instruments and Methods of Microscopy." *Perspectives on Science* 9 (2001): 126–45.

Schirrmacher, Arne. "Das leere Atom. Instrumente, Experimente und Vorstellungen zur Atomstruktur um 1903." In *Circa 1903: Artefakte in der Gründungszeit des Deutschen Museums*, edited by Ulf Hashagen, Oskar Blumtritt, and Helmuth Trischler. München: Deutsches Museum, 2003, 127–52.

Schirrmacher, Arne. "Der lange Weg zum neuen Bild des Atoms. Ein Problemfall der Wissenschaftsvermittlung von der Jahrhundertwende bis in die Zeit der Weimarer Republik." In *Wissenschaft und Öffentlichkeit als Ressourcen füreinander. Studien zur Wissenschaftsgeschichte im 20. Jahrhundert*, edited by Sibylla Nikolow and Arne Schirrmacher, 39–73. Frankfurt and New York: Campus, 2007.

Sichau, Christian. "Atome. Eine lange Geschichte." In *Abenteuer der Erkenntnis. Albert Einstein und die Physik des 20. Jahrhunderts*, edited by Alto Brachner, Gerhard Hartl, and Christian Sichau. München: Deutsches Museum, 2005, 142–51.

Sieverking, H. "Sichtbarmachung der Moleküle nach Laue und Wilson." *Kosmos* 10 (1913): 265–68.

Vinnikov, Victor. "We Shall Know: Hilbert's Apology." *Mathematical Intelligencer* 21 (1999): 42–46.

von Miller, Oskar, to Peter Debye. January 16, 1920. Deutsches Museum Archives, object accession folder 47887.

Zsigmondy, Richard, and Richard Siedentopf. "Über Sichtbarmachung und Größenbestimmung ultramikroskopischer Teilchen, mit besonderer Anwendung auf Goldrubingläser." *Annalen der Physik* 10 (1903): 1–39.

CHAPTER 9

Instrument Museums and Collections

Klaus Staubermann

ANY HISTORY of scientific instruments is a history of objects and of practice. The synthesis of object and practice is manifested in the instruments' functioning—both technical and cultural—about which this volume offers various narratives. By concentrating on the process of designing the device, we learn about the techniques involved in constructing it. By tracing the instrument as it was treated by its users, we learn how the device was transformed and interpreted. We learn how the instrument was employed, how it served in the evolution of scientific disciplines, what rituals it helped to form, and its wider societal significance. No history of scientific instruments can be written without reference to the objects themselves. Hence, museums and collections are the most crucial resource for any historian working on scientific instruments. The following list presents a number of such museums and collections, providing both readers and historians with references for future studies. The list is categorized in this order: alphabetical by location, alphabetical by name of institution. The list is by nature incomplete. Further information on instrument collections may be found on the UMAC/ICOM collection database: www.publicus.culture.hu-berlin.de/collections/

Australia

Macleay Museum (University of Sydney)—www.usyd.edu.au/museums/

The Macleay Museum's scientific instrument collection contains more than 1,000 items, covering microscopy, electrical, and chemical apparatus, glassware, weights and measures, surveying, navigation, drawing, and meteorological instruments.

Belgium

Museum voor des Geschiedenis van de Wetenschappen (University of Ghent)—www.mhsgent.ugent.be

Belgium's largest collection of scientific instruments features, among others, optical and electrical instruments, including those of Joseph Plateau and August Kekulé.

Canada

Canada Science & Technology Museum (Ottawa)—www.sciencetech.technomuses.ca

The Canadian Science & Technology Museum holds Canada's largest general collection of scientific apparatus dating from the seventeenth century to the present. It covers more than 5,000 artifacts, including documentation such as manuals, trade literature, and photos.

Denmark

Steno Museum (Aarhus)—www.stenomuseet.dk

Denmark's History for Science and Medicine Museum's scientific instrument collection includes geodetic and smaller astronomical instruments as well as some larger telescopes from around 1860. It covers the history of electromagnetism and telegraphy. Modern science covers spectroscopes, cloud chambers, transistors, and one of the first computers built in Denmark in the late 1950s.

Hauch Physiske Cabinet (Soroe)—www.awhauch.dk

The Hauch collection, displayed in its original eighteenth-century cabinet, covers the areas of statics and navigation, Newtonian mechanics, chemistry and gases, water machines, electricity, light, and heat and pressure.

Estonia

Tartu Ùlikooli Ajaloo Museum (University of Tartu)—www.ut.ee/ajaloomuuseum/museum.html

Estonia's largest scientific instrument collection includes makers such as Zeiss, Leitz, and Schick (microscopes); Sartorius (scales); Askana (measuring devices); Dolland (optic devices); Kohl (laboratory equipment); Meissner (areometers); Geissler (devices for measuring temperature), and Schmidt & Haensch (spectroscopes). The oldest object on display is a thirteenth-century celestial globe.

France

Musée des Arts et Métiers–CNAM (Paris)—www.arts-et-metiers.net

The collection at the Musée des Arts et Métiers comprises some original demonstration machines used in eighteenth-century physics cabinets; weighing scales, gasometers, and calorimeters (made by Lavoisier, Fortin, and Meusnier and used for experiments on water synthesis and metal combustion); the first programming machines (Pascal's calculating machines, Hollerith's machine, the first computers); and others (Foucault's pendulum, machines for producing electricity, automaton-musicians).

Germany

Arithmeum (Bonn)—www.arithmeum.uni-bonn.de

The collection of historical mechanical calculating machines comprises more than 1,200 pieces and is the largest in the world, covering a period from the first beginnings in the seventeenth century to their demise some two decades ago.

Astronomisch-Physikalisches Kabinett (Kassel)—www.museum-kassel.de

This collection founded by the Count of Hesse covers scientific instruments from the late renaissance till the early industrial revolution and includes mechanical globes, astronomical clocks, vacuum pumps, microscopes, electrostatic machines, quadrants, and early calculating machines.

Deutsches Museum (Munich)—www.deutsches-museum.de

With 28,000 displayed artifacts covering about fifty areas of science and technology the Deutsches Museum is the world's biggest museum of its kind. Its collection of scientific instruments has its origin in 2,000 instruments from the Bavarian Academy of Science donated to the Deutsches Museum's founder Otto von Miller in 1903. Among the many displays dedicated to scientific instruments is a re-created instrument-making workshop and hall of fame dedicated to Joseph Fraunhofer.

Mathematisch-Physikalischer Salon (Dresden)—www.skd-dresden.de

The Mathematisch-Physikalischer Salon houses a collection of historical timepieces and instruments. Terrestrial and celestial globes are on view alongside optical, astronomical, and geodetic devices of the sixteenth to nineteenth centuries, and the Salon's compendium also includes instruments for calculating and drawing as well as determining length, dimensions, temperature, and air pressure.

Italy

Istituto e Museo di Storia Della Scienza (Florence)—www.imss.fi.it

The Istituto e Museo di Storia della Scienza is heir to a tradition of five centuries of scientific collecting, which has its origins in the scientific instruments of the Medici and Lorraine families. The museum's Renaissance instruments include Galileo's geometric and military compass, armed loadstone, two telescopes, and the objective lens of the telescope with which Galileo discovered the Jupiter satellites.

Museo di Storia della Fisica (University of Padua)—www.musei.unipd.it

In 1740, Giovanni Poleni inaugurated his Teatro di Filosofia Sperimentale, the first physics cabinet in an Italian university. Poleni's successors continued to acquire instruments for the Teatro. Nowadays, the museum houses several thousand instruments that cover the history of physics and mathematics from the sixteenth century onward.

Netherlands

Museum Boerhaave (Leiden)—www.museumboerhaave.nl

The Boerhaave's earliest objects date from the middle of the sixteenth century. The collections include many instruments from the Netherlands' golden age, such as Willem Blaeu's giant quadrant, microscopes by Antoni van Leeuwenhoek, and clocks by Christiaan Huygens, as well as his planetarium and telescope. Eighteenth-century instruments encompass those by Gravesande and Van Musschenbroek. Twentieth-century devices relate to Dutch

Nobel Prize laureates such as Van't Hoff, Lorentz, Zeeman, Van der Waals, Kamerlingh Onnes, and Willem Einthoven.

Nederlands Scheepvaartmuseum (Amsterdam)—www.scheepvaartmuseum.nl
The collection consists of 850 items, most of them navigation devices for the determination of position, speed, depth, and direction as well as radio- and radar technology, oceanographic apparatus, chronometers, and others.

Teylers Museum (Haarlem)—www.teylersmuseum.nl/index_flash.html
Founded in 1784 by Pieter Teyler van der Hulst, the Teylers Museum is considered the oldest in the Netherlands. Still situated in their original setting, more than 1,200 instruments are on display or in storage. Exceptional pieces are Marinus van Marum's giant electrostatic generator dating 1784, a large collection of Lyeden jars, and celestial and terrestrial globes by George Adams.

Universiteitsmuseum (University of Utrecht)—www.uu.nl/uupublish/4192main.html
The collection of scientific instruments lists several significant pieces, such as Van Leeuwenhoek's only early microscope with a blown lens, a Van Musschenbroek air pump, and a lens used by Christiaan Huygens for the discovery of Saturn's satellite Titan. Two sections of the collection of international significance are microscopes (approximately 1,000 objects) and physiology (associated with F. C. Donders).

Poland

Muzeum Uniwersiytetu Jagiellonskiego (University of Kracow)—www3.uj.edu.pl/Muzeum/index.en.html
Poland's most important collection of scientific instruments is comprised of more than 2,300 objects. Among these objects are medieval astronomical instruments, including a Moorish astrolabe from 1054; fifty-five globes, some of them made by Mercator and Blaeu and Abel and Klinger; quadrants by Ramsden; and telescopes by Dollond. Further outstanding sections of the collection cover microscopes, drawing tools, and calculating aids.

Portugal

Museu de Física (University of Coimbra)—www.museu.fis.uc.pt/indexi.htm
Thanks to its unique characteristics, Portugal's most prominent collection of scientific instruments is among the most notable and rare in the world. Most of these scientific and didactic instruments date from the eighteenth and nineteenth centuries. The patrimony consists exclusively of instruments used in the Physics Cabinet of the University of Coimbra since its origin in 1772. Many of the instruments are still kept in their original cabinets.

Switzerland

Musée d'Histoire des Sciences (Geneva)—www.ville-ge.ch/culture/mhs
The Museum of the History of Science is the only museum of its kind in Switzerland. Its collections are composed of scientific instruments including microscopes, barometers,

sundials, astrolabes, as well as books and documents donated by the families of Geneva's scientists and scholars.

United Kingdom

Museum of the History of Science (University of Oxford)—www.mhs.ox.ac.uk

The objects—of which there are approximately 10,000—cover almost all aspects of the history of science, from antiquity to the early twentieth century. Particular strengths include the collections of astrolabes, sundials, quadrants, general early mathematical instruments (including those used for surveying, drawing, calculating, astronomy, and navigation), and optical instruments (including microscopes, telescopes, and cameras), together with apparatus associated with chemistry, natural philosophy, and medicine.

National Maritime Museum, with the Old Royal Observatory, Greenwich (London)—www.nmm.ac.uk

The National Maritime Museum's collections contain over 2 million items relating to seafaring, navigation, astronomy, and time measurement, many of them scientific instruments. Among the most important instruments are the timekeepers by John Harrison, the large transit circle by George Airy, and several telescopes made by William and Alexander Herschel.

Science Museum (London)—www.sciencemuseum.org.uk

The Science Museum is Great Britain's biggest science museum. It holds more than 200,000 objects in its collections. Outstanding objects are William Herschel's forty-foot telescope mirror, the King George III collection of early demonstration instruments, Charles Babbage's Analytical Engine, conceived in 1834, and J. J. Thomson's Nobel Prizewinning experimental apparatus.

National Museums of Scotland (Edinburgh)—www.nms.ac.uk

Science collections at the National Museum include scientific instruments, measurement and precision apparatus, Scottish lighthouse optics, and apparatus associated with pure and applied research in modern science. Strengths of the collection are instruments made during the Scottish Enlightment and Industrial Revolution.

Whipple Museum of the History of Science (University of Cambridge)—www.hps.cam.ac.uk/whipple/index.html

The museum's holdings are particularly strong in material dating from the seventeenth to the nineteenth centuries, especially objects produced by English instrument-makers. The collection also contains objects dating from the medieval period to the present day. Instruments of astronomy, navigation, surveying, drawing, and calculating are well-represented, as are sundials, mathematical instruments, and early electrical apparatus.

United States of America

Adler Planetarium & Astronomy Museum (Chicago)—www.adlerplanetarium.org

Housed in the Adler's Webster Institute for the History of Astronomy, the Scientific Instrument Collection today contains about 2,000 instruments and models from the twelfth through the twentieth centuries. Representing many types of astronomical instruments, it is the largest collection of such material in the Western Hemisphere and one of the most significant in the world.

Bakken Library and Museum (Minneapolis)—www.thebakken.org

The instrument collection includes approximately 2,500 objects. The focus is on the historical role of electricity and magnetism in life sciences and medicine. This encompasses instruments of electricity, electrophysiology, and electrotherapeutics, during the eighteenth, nineteenth, and early twentieth centuries.

Harvard University Collection of Historical Scientific Instruments (Harvard University)— www.fas.harvard.edu/~hsdept/chsi.html

The Collection of Historical Scientific Instruments, which was established in 1949 to preserve this apparatus as a resource for teaching and research in the history of science and technology, now contains over 20,000 objects dating from about 1400 to the present. A broad range of scientific disciplines are represented, including astronomy, navigation, horology, surveying, geology, calculating, physics, biology, medicine, psychology, electricity, and communication.

National Museum of American History (Smithsonian Institution, Washington, D.C.)— www.americanhistory.si.edu

The physical sciences collection includes apparatus of astronomy, chemistry, classical physics, meteorology, navigation, and surveying—some for research, some for education, and some for practical purposes. Trade literature supplements the collection. The modern physics collection includes instruments, experimental apparatus, and other objects relating to radioactivity; the production, acceleration, and detection of subatomic particles; cryogenics; the phenomena dealt with by quantum and relativity theory; atomic beams; atomic frequency standards; and other branches of contemporary physics.

Select Bibliography

Alder, Ken. *The Measure of all Things: The Seven-Year Odyssey and Hidden Error that Transformed the World*. New York and London: Free Press, 2002.

Anderson, R. G. W., J. A, Bennett, and W. F. Ryan, eds. *Making Instruments Count: Essays on Historical Scientific Instruments Presented to Gerard L'Estrange Turner*. Aldershot: Variorum, 1993.

Baird, Davis. *Thing Knowledge: A Philosophy of Scientific Instruments*. Berkeley: University of California Press, 2004.

Bourguet, Marie-Noëlle, Christian Licoppe, and H. Otto Sibum. *Instruments, Travel and Science: Itineraries of Precision from the Seventeenth to the Twentieth Century*. London and New York: Routledge, 2002.

Bud, Robert, and Susan Cozzens, eds. *Invisible Connections: Instruments, Institutions and Science*. Bellingham, WA: SPIE Optical Engineering Press, 1992.

Bud, Robert, and Deborah Jean Warner, eds. *Instruments of Science: An Historical Encyclopedia*. New York and London: Garland Publishing, 1998.

Collins, Harry, and Martin Kusch. *The Shape of Actions: What Humans and Machines Can Do*. Cambridge, MA: MIT Press, 1998.

Dorikens, Maurice, ed. *Scientific Instruments and Museums*. Proceedings of the 20th International Congress of the History of Science, vol. 26. Turnhout: Brepols, 2002.

Grob, Bart, and Hans Hooijmaijers, eds. *Who Needs Scientific Instruments? Conference on Scientific Instruments and their Users*. Leiden: Museum Boerhaave, 2006.

Hankins, Thomas L., and Robert J. Silverman. *Instruments and the Imagination*. Princeton, NJ: Princeton University Press, 1995.

Joerges, Bernward, and Terry Shinn. *Instrumentation Between Science, State and Industry*. Dordrecht: Kluwer, 2000.

Lüthy, Christoph. "Museum Spaces and Spaces of Science. Reflections on the Explanatory Possibilities of History of Science Collections." *Nuncius* 20 (February 2005): 415–29.

Meinel, Christoph. *Instrument-Experiment: Historische Studien*. Berlin: Verlag für Geschichte der Naturwissenschaften und der Technik, 2000.

Morris, Peter J. T. *From Classical to Modern Chemistry: The Instrumental Revolution*. Cambridge, UK: Royal Society of Chemistry in association with the Science Museum, 2002.

Shapin, Steven, and Simon Schaffer. *Leviathan and the Air-Pump: Hobbes, Boyle and the Experimental Life.* Princeton, NJ: Princeton University Press, 1985.

Staubermann, Klaus. *Astronomers at Work: A Study of the Replicability of Nineteenth Century Astronomical Practice.* Frankfurt: Harri Deutsch, 2007.

Thirring, H. *Die Idee der Relativitatstheorie.* Berlin: Springer, 1921.

Van Helden, Albert, and Thomas L. Hankins, eds. "Instruments." *Osiris* 9 (1994): 1–250.

Warner, Deborah Jean. "What is a Scientific Instrument, When Did it Become One, and Why?" *British Journal for the History for the History of Science* 23 (1990): 83–93.

Index

About the Contributors

Peter Heering is senior lecturer at the Physics Department of the Carl-von-Ossietzky University in Oldenburg, Germany. His research focuses on the analysis of experimental practice with the replication method. Actual projects attempt to analyze changes in experimental practice and to investigate the relation between research experiments and teaching demonstrations.

Sean F. Johnston is Reader in history of science and technology at the University of Glasgow. His research focuses on the emergence of new disciplines and technical professions, with books on the history and sociology of chemical engineering, spectroscopy, light measurement, and, most recently, holography.

Peter J. T. Morris studied chemistry at Oxford University and did his doctorate on IG Farben. He was the first Royal Society–British Academy research fellow before joining the Science Museum in 1991. He is currently the principal curator of science. He was given the American Chemical Society's Edelstein Award in the History of Chemistry in 2006.

Arne Schirrmacher teaches both at the universities of Munich and Augsburg and currently is a visiting scholar at the Max Planck Institute for the History of Science in Berlin. His main research interests lie in the history of science, twentieth-century science communication, and the rationality of experiment and theory.

Christian Sichau, head of exhibitions development at the Experimenta Science Center in Heilbronn, received a PhD in physics from the University of Oldenburg, studied at King's College London, and worked at the Technisches Museum in Vienna. Major research interests include history of physics and scientific instruments in the nineteenth and early twentieth centuries.

Klaus Staubermann is principal curator of technology at National Museums Scotland in Edinburgh and acts as the Museums' science communication coordinator. He is interested in historic practices and has published on the reconstruction of historic observations, experiments, and instruments.

Deborah Jean Warner, curator of the physical sciences collections at the National Museum of American History, has written widely on the history of scientific and mathematical instruments.

Jane Wess has worked at the Science Museum since 1979. She specializes in eighteenth-century natural philosophy, the history of mathematical instruments, and, more recently, classical physics. With Alan Morton she wrote *Public and Private Science*, published in 1993, and with Silke Ackermann contributed to the Enlightenment gallery at the British Museum.